Y0-FZM-207

UNDERSTANDING HISTOCHEMISTRY:
Selection, Evaluation and Design of Biological Stains

Ellis Horwood books in the
BIOLOGICAL SCIENCES
General Editor: Dr ALAN WISEMAN, University of Surrey, Guildford

Series in
BIOCHEMISTRY AND BIOTECHNOLOGY

Series Editor: Dr ALAN WISEMAN, Senior Lecturer in the Division of Biochemistry, University of Surrey, Guildford

Aitken, J.	**Handbook of Enzyme Active Site Identification***
Ambrose, E.J.	**The Nature and Origin of the Biological World**
Austin, B. & Brown, C.M.	**Microbial Biotechnology: Freshwater and Marine Environments**
Berkeley, R.C.W., *et al.*	**Microbial Adhesion to Surfaces**
Blackburn, F. & Knapp, J.S.	**Agricultural Microbiology***
Bowen, W.R.	**Membrane Separation Processes***
Bubel, A. & Fitzsimons, C.	**Atlas of Cell Biology***
Butt, W.R.	**Topics in Hormone Chemistry Vol. 1**
Corkill, J.A.	**Clinical Biochemistry: The Analysis of Biologically-important Compounds and Drugs***
Crabbe, M.J.C.	**Enzyme Biotechnology and Protein Engineering**
Dean, A.C.R., *et al.*	**Continuous Culture Vols. 6 & 8**
Dolly, J.O.	**Neurotoxins in Neurochemistry***
Dunnill, P., Wiseman, A. & Blakeborough, N.	**Enzymic and Non-enzymic Catalysis**
Espinal, J.	**Understanding Insulin Action***
Eyzaguirre, J.	**Chemical Modification of Enzymes**
Eyzaguirre, J.	**Human Biochemistry and Molecular Biology***
Fish, N.M.	**Computer Applications in Fermentation Technology***
Horobin, R.W.	**Understanding Histochemistry: Selection, Evaluation and Design of Biological Stains**
Inoki, R., Kudo T. & Olgart, L.M.	**Dynamic Aspects of Dental Pulp***
Kennedy, J.F. & White, C.A.	**Cellulose and its Derivatives**
Kennedy, J.F. & White, C.A.	**Bioactive Carbohydrates**
Kricka, L.J. & Clark, P.M.S.	**Biochemistry of Alcohol and Alcoholism**
Palmer, T.	**Understanding Enzymes, 2nd Edition**
Pasteur, N., *et al.*	**Isoenzyme Detection: Methods and Principles***
Reid, E.	**Methodological Surveys in Biochemistry Vols. 6–11**
Reizer, J. & Peterkofsky, A.	**Sugar Transport and Metabolism in Gram-positive Bacteria**
Roe, F.J.C.	**Microbiological Standardisation of Laboratory Animals**
Russell, N. J.	**Microbes and Temperature***
Russell, N.J. & Gacesa, P.	**Pseudomonas Infection and Alginates: Structures, Properties and Functions in Pathogenesis***
Sammes, P.G.	**Topics in Antibiotic Chemistry Vols. 1–6**
Scragg, A.H.	**Biotechnology for Engineers***
Sikyta, B. & Moss, M. O.	**Methods in Industrial Microbiology**
Sluyser, M. & Voûte, P.A.	**Neuroblastoma: New Approaches to Diagonosis and Therapy***
Verrall, M.S.	**Discovery and Isolation of Microbial Products**
Verrall, M.S. & Hudson, M.J.	**Separations for Biotechnology**
Webb, C. & Mavituna, F.	**Process Possibilities in Plant and Animal Cell Cultures**
Wiseman, A.	**Handbook of Enzyme Biotechnology, 2nd Edition**
Wiseman, A.	**Topics in Enzyme and Fermentation Biotechnology Vols. 1–10**
Wiseman, A.	**Enzyme Induction, Mutagen Activation and Carcinogen Testing in Yeast**

* *In preparation*

UNDERSTANDING HISTOCHEMISTRY:
Selection, Evaluation and Design of Biological Stains

R. W. HOROBIN
Department of Anatomy and Cell Biology
University of Sheffield

ELLIS HORWOOD LIMITED
Publishers · Chichester

Halsted Press: a division of
JOHN WILEY & SONS
New York · Chichester · Brisbane · Toronto

First published in 1988 by
ELLIS HORWOOD LIMITED
Market Cross House, Cooper Street,
Chichester, West Sussex, PO19 1EB, England
The publisher's colophon is reproduced from James Gillison's drawing of the ancient Market Cross, Chichester.

QH
613
H67
1988

Distributors:
Australia and New Zealand:
JACARANDA WILEY LIMITED
GPO Box 859, Brisbane, Queensland 4001, Australia
Canada:
JOHN WILEY & SONS CANADA LIMITED
22 Worcester Road, Rexdale, Ontario, Canada
Europe and Africa:
JOHN WILEY & SONS LIMITED
Baffins Lane, Chichester, West Sussex, England
North and South America and the rest of the world:
Halsted Press: a division of
JOHN WILEY & SONS
605 Third Avenue, New York, NY 10158, USA
South-East Asia
JOHN WILEY & SONS (SEA) PTE LIMITED
37 Jalan Pemimpin # 05–04
Block B, Union Industrial Building, Singapore 2057
Indian Subcontinent
WILEY EASTERN LIMITED
4835/24 Ansari Road
Daryaganj, New Delhi 110002, India

© 1988 R. W. Horobin/Ellis Horwood Limited

British Library Cataloguing in Publication Data
Horobin, Richard W.
Understanding histochemistry.
1. Histochemistry
I. Title
574.8′212

Library of Congress CIP available

ISBN 0–85312–678–X (Ellis Horwood Limited)
ISBN 0–470–21060–5 (Halsted Press)

Phototypeset in Times by Ellis Horwood Limited
Printed in Great Britain by The Camelot Press, Southampton

COPYRIGHT NOTICE
All Rights Reserved. No part of this publication may be reproduced, stored in a retrieval system, or transmitted, in any form or by any means, electronic, mechanical, photocopying, recording or otherwise, without the permission of Ellis Horwood Limited, Market Cross House, Cooper Street, Chichester, West Sussex, England.

Table of contents

Appendices

Part I

Introductory information

I’ll skip/ Upon the trees of Paradise

R.L.S.

1

How to use this book

I envisage you, the reader, as wanting to use this book to achieve one or more of the following objectives:

(1) To discover how the biological problems you are concerned with might be tackled using histochemical methods.
(2) To find specific items of histochemical information: perhaps concerning a staining technique or reagent; or a concept, or an item of nomenclature.
(3) To obtain a general technical background to formal educational coursework in biology, histopathology and so on.
(4) To seek enlightenment concerning the scientific principles of staining biological specimens, prior to microscopic study.

Those readers with biological problems without obvious technical answers should go directly to Chapter 9 'Selection and design of staining methods'. If desiring information concerning particular processes or known techniques, glance through the Table of Contents before delving into the Index. Reading Chapters 1–7 and Chapter 9 is recommended for most students. However, such readers should first look carefully at the Table of Contents, to obtain an overall view of the material presented. Whilst appreciating that seekers of enlightenment are often browsers, I do recommend scanning the Table of Contents to them also.

Readers falling into categories (3) and (4) will profit from, and all readers will enjoy, tackling the 'tests of understanding' found at the ends of most chapters. These sections may well demand information from other chapters, or indeed from other books, for their solution. Many of the questions posed have multiple valid answers!

All readers should note that although citations of books, review articles and research papers do provide a certain amount of small-print material, the flavour of this book is pedagogical not scholarly. The book does provide gateways into the histochemical and related literatures; it does not, however, attempt to do justice to the priorities of scientific discovery or technical innovation.

Finally note that my vision of this book's role is that its readers may come to understand the nature of useful histochemical questions, rather than merely find 'answers'.

2

The whats, whys and hows of histochemistry

2.1 WHAT IS HISTOCHEMISTRY?

To assist the microscopic study of the geometry, content and function of biological materials, histochemical staining is often carried out. Staining is essentially a process of visual labelling, with the nature of the label being extremely varied. For instance parts of the material under study may be coloured, or blackened, in some way. The colour may be due to absorption or emission (fluorescence) or scattering of light. On the other hand a stain used in electron microscopy will normally be electron scattering. Occasionally, in both light and electron microscopy, the identifying feature of a stain is neither colour nor electron density but its unusual shape.

The question 'What is histochemistry?' also has a terminological aspect. Much the same range of manipulative methods, involving similar physicochemical processes, are variously known as biochemical microtechnique, biochemical morphology, biological staining, cytochemistry, histochemistry, histopathologic technique—and doubtless more. The present book, for simplicity and to conform with a majority of other writers, most commonly uses the term 'histochemistry'.

2.2 WHY CARRY OUT HISTOCHEMICAL STAINING?

The fundamental reason for such staining is simple enough. Chloroplasts, erythrocytes, pigment granules and elastic fibres apart, few biological entities are brightly coloured or strongly fluorescent. The need to identify natively invisible entities using standard microscopes is commonplace: hence histochemical staining. Of course viewing colourless objects with phase contrast or interference microscopy can be informative and is at times also carried out.

What kinds of information can be provided by histochemical investigation? Briefly, histochemistry provides information concerning *what* and *where* something is, and sometimes *how much* of it is present or *how many* of them there are.

The essence of microscope-based methods for investigating biological materials and processes is the provision of information on where something is. Thus, when using the microscope, all 'what is it?' 'how much?' and 'how many?' questions are placed in a morphological context. In a simple case, staining discloses where in an organism or cell or organelle a particular substance is to be found.

However, 'What is it?' questions do vary in focus. Attention may indeed be focused on *chemical entities*. It may be necessary to demonstrate the presence of a class of compounds, such as neutral lipids; or an individual compound, for instance succinate dehydrogenase; or perhaps chemical fragments, say calcium ions.

On the other hand the focus of study may be on *biological objects,* which are characterized by their biological significance rather than by chemical composition. Objects of this type include whole creatures, such as Gram-positive bacteria; particular tissues, such as muscle; cell types, plasma cells for instance; organelles, such as lysosomes; or even more fragmentary objects such as antigens.

It may be biological *processes* which are studied, and not objects at all. Axonal transport, membrane fluidity, phagocytosis, and secretion are a few of the processes whose investigation has been aided by histochemistry. Indeed the question 'Are these cells alive?' is sometimes conveniently answered by a staining method.

Another non-chemical property whose study is often assisted by histochemical staining is morphology. Shapes and distribution of cells and organelles may be studied. For examples changes in form and situation of macrophages and their constituent lysosomes during the process of phagocytosis. The existence of spaces within cells or tissues can be investigated; epithelial intercellular clefts are often explored in this way. Connectivity, too, can be probed by staining methods, as in the study of neuronal pathways.

Tackling 'How much?' questions is often described as quantitative histochemistry. Oddly, tackling 'How many?' questions is not so regarded; or is regarded as morphometry rather than histochemistry. Though 'How much?' questions have been answered histochemically for many chemical entities, from amino acid residues to zinc, it is only DNA and certain enzymes which are assayed at all frequently. Indeed it is perhaps the 'How many?' questions which are asked and answered most commonly. Think of all those automated systems in haematology departments telling us how many basophils or monocytes there are in a blood sample.

A notable feature of these examples is the enormous size range of the objects studied. Essentially the same staining reactions can detect entities as small as molecular fragments, or as large as complete organisms. Indeed staining methods are occasionally used macroscopically, to demonstrate cardiac infarct in post mortem muscle samples for instance.

2.3 HOW IS HISTOCHEMICAL STAINING ACHIEVED?

2.3.1 Introduction

First we must specify what is required to permit effective microscopic observation of samples of biological materials. Several distinct issues may be distinguished, as follows.

The content and structure of live cells and tissues must be stabilized. Otherwise, after staining and processing, there may be little left to observe! Most unstabilized biological materials are extremely labile. Autolysis may occur, or micro-organisms may eat the cells. Processing and staining procedures can extract many of the components of cells and tissues; and also give rise to mechanical, osmotic, and chemical damage.

Useful microscopic observation requires specimens of appropriate thickness. Too thin, and optical contrast may be insufficient, and interrelation of constituent parts unobservable. Too thick, and an excess of detail may confuse.

Special techniques are needed to manipulate cell layers only micrometers thick prior to, during, and after staining.

Obviously the material must have sufficient visual contrast to allow effective observation: consequently the material is usually stained.

The thin, stained layers prepared for microscopic observation are mechanically fragile, and readily fade. Furthermore hungry micro-organisms, and even insects, abound. Consequently the coloured layers must be carefully protected.

Ways of dealing with these problems will now be discussed in more detail. A diagrammatic summary of problems and of common responses, is found in Fig. 2.1.

Imagine you wish to study a collections of cells

It will usually be necessary to	. . . consequently you
STABILIZE the cells	fix; or use membranes or colloid stabilizers when staining
obtain THIN LAYERS	smear or dab; or section after freezing or embedding
MANIPULATE these fragile layers without causing damage	lay thin layers on glass slides, or on grids, or plastic film
make the cells VISUALLY CONTRASTY	stain: usually the thin layers, sometime the whole cells
STABILIZE the thin, contrasty preparation	mount in oil or a mounting medium, usually under a cover slip

You will now have preparations suitable for observing in a light microscope

Fig. 2.1 — Procedures involved in preparing biological specimens for microscopic study.

2.3.2 Stabilization of biological materials

It has long been appreciated by microscopists that in the absence of stabilization, stained preparations differ grossly in shape, size and chemical content from their live originals. Thus clearing agents extract lipids, aqueous stains extract glycogen, and paraffin embedding causes substantial shrinkage.

Traditionally, microscopists stabilized cells and their contents by *fixation*. This entails conversion of soluble native constituents into insoluble derivatives, by treatment with some fixative agent. Such agents include organic solvents; chemically reactive compounds, both organic and inorganic compounds; and intangibles such as heat. Though fixative agents are varied, most specimens are insolubilized by one or more of a limited range of processes: cross-linking, denaturation, formation of insoluble complexes, and mechanical trapping. Examples of these are given in Table 2.1.

Table 2.1 — Fixation of widely occurring biological compounds

Compounds insolubilized	Fixative mechanism	Fixatives acting in this way	Comments
proteins	crosslinking	glutaraldehyde	via NH_2 groups
unsaturated lipids		osmium tetroxide & $K_2Cr_2O_7$	via C=C groups
nucleic acids & proteins	denaturation	most fixatives are denaturants, whatever else they do	involves disruption of complex molecular structures
anionic GAGs (glycosaminoglycans)	insoluble complex formation	hydrophobic cations, e.g. cetyl pyridinium	formation of insoluble salts
metal ions		Sulphide ions	
unreactive compounds, e.g. glycogen & insulin	trapping	any effective denaturant of proteins	compounds held inside cages of insolubilized proteins

Although traditional, fixation is far from an ideal stabilizing strategy. Fixatives themselves can, and usually do, give rise to substantial losses of content and changes in form; as shown in Fig. 2.2. Moreover fixation inevitably changes the native chemistry of retained compounds. This is especially troublesome when studying enzymic activity or protein antigenicity, since protein conformations are altered during denaturation. As a result, the possibility of stabilization by non-destructive fixation, or indeed by procedures other than fixation, have been investigated.

One much used strategy is the careful matching of fixation procedure to sample and to overall objective. For instance if enzymic activity is to be studied, prolonged exposure of the sample to effective protein denaturants such as glutaraldehyde must be avoided. However, if preservation of the details of intracellular structure is important, then glutaraldehyde plus osmium tetroxide will be a fixative of choice. There is no single ideal fixative, the choice of optimum regime is achieved by juggling fixative identity (glutaraldehyde or acetone or acrolein and so on) and fixative conditions (which pH, temperature, concentration, osmolarity, time? Perfusion or immersion? Liquid or vapour?). Though there are general principles (such as: to fix

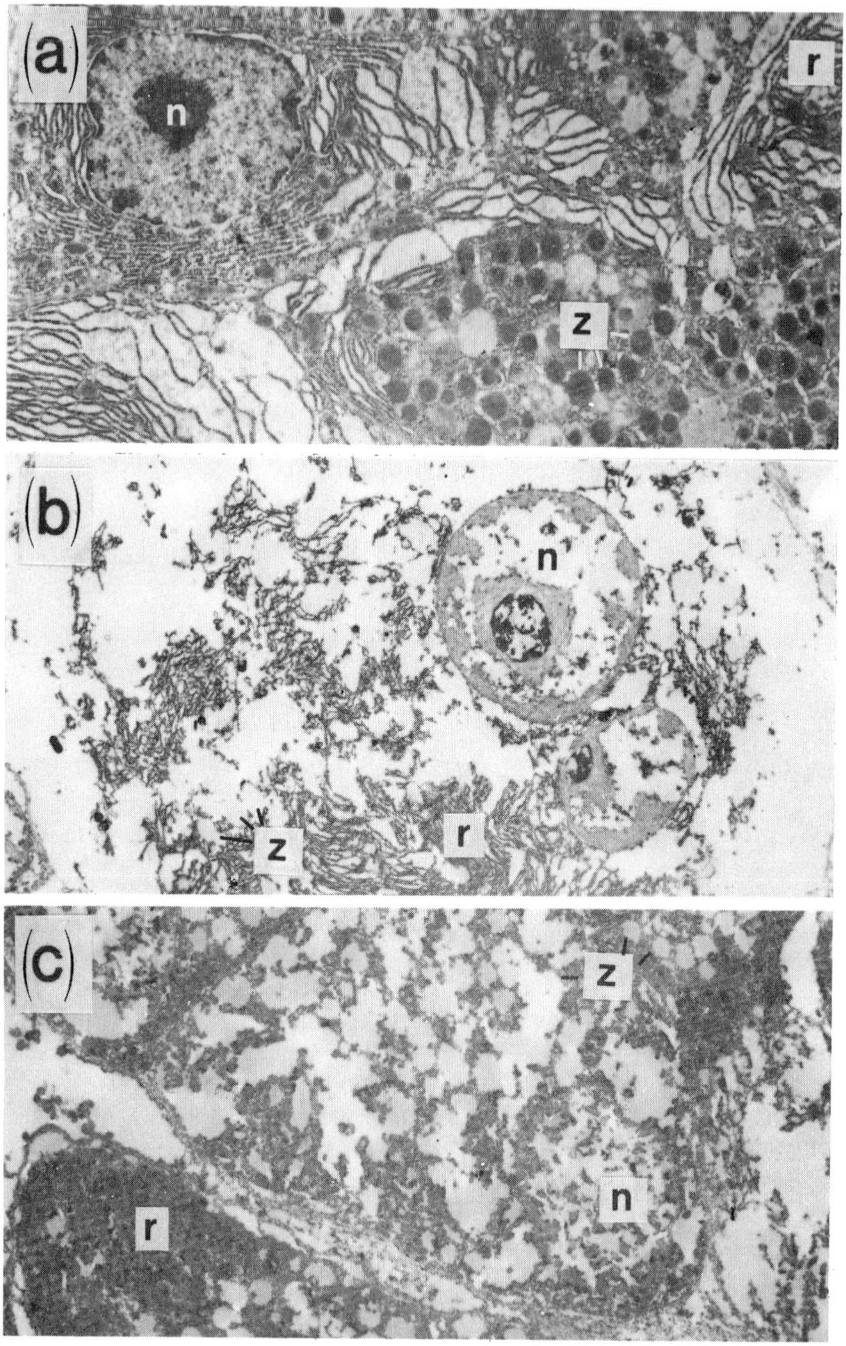

Fig. 2.2 — Effects of various fixative agents on cell structure. Micrographs (a), (b), and (c) resulted from the use of neutral formalin, Carnoy's fluid, and alcohol respectively to fix mouse pancreas. Structural features are indicated as follows: **n** are nuclei, **r** is rough endoplasmic reticulum, and **z** are zymogen granules (or, in (b) and (c), the *in vivo* site of the granules). Magnification is 4250 times.

glycogen, avoid aqueous solutions), optimization is achieved empirically. Nevertheless, remarkably successful fixation schedules can be arrived at, given sufficient skill, time, and good fortune.

Fixation-avoidance strategies of stabilization are themselves varied. The most widely used way of evading the traumas of fixation, and indeed those of embedding, is to use cryostat sections. Disadvantages include the difficulty of achieving good morphological preservation, and the fact that losses of content are not eliminated but only deferred. For instance glycogen present in cryostat sections is rapidly leached out into aqueous solutions of periodic acid and Schiff's reagent when the Periodic acid-Schiff stain is carried out. Hence attention has focused on preventing losses into staining solutions. One strategy finding increasing favour is to separate the cell contents from the solvent. Sections may be coated with a thin semipermeable membrane or, an equivalent in molecular terms, immersed in staining solutions containing high concentrations of inert polymers (colloid stabilizers). Both procedures trap tissue macromolecules within the section, whilst allowing free entry of small stain molecules. Neither process directly prevents losses of small molecules, or damage to morphology.

2.3.3 Obtaining thin layers

Though some specimens are naturally thin, most are not and hence require thinning. How this is achieved varies, since biological materials may be fluids, or solids of very varied textures.

Inately thin materials can be merely draped over a microscope slide: think of viewing the circulation of blood through a frog's foot web; or looking at onion scale leaves.

Fluids are a special case of materials which do need forming into thin layers. Some are easily observed merely by lowering a cover slip onto a drop of the fluid on a microscope slide; the traditional way of observing the creatures swarming in murky pond water. The most studied fluid, however, is blood. Layers one cell thick are readily prepared by smearing or by using a spinner or an equivalent device. Cells in other body fluids, such as urine, may need concentration. This can be achieved by filtration which gives rise to a layer of cells on the surface of the filter, or by centrifugation.

Solid specimens vary greatly in consistency, with some being rather soft. Thus bone marrow, after puddling in a drop of buffer, can be smeared into a film. Although lymph nodes, spinal cord or brain are more rigid, cells from these structures are readily deposited onto a microscope slide by dabbing or smearing. Even solid tissues may have detachable surface layers. Cervical smears are obtained by wiping the epithelium with a swab, then smearing the cells from the swab onto a microscope slide. Epidermal cells can be removed from the skin following their adhesion to sticky tape.

Most animal and plant tissues of moderate rigidity are, however, formed into thin layers by microtomy. This is the slicing the tissue by a mechanical device (a microtome) with a cutting edge, traditionally of steel though increasingly often of fractured glass, and occasionally of tungsten carbide or even diamond. Sectioning of materials such as liver or leaves usually requires specimens to be mechanically supported by being embedded within a solid matrix. Embedding involves the specimen being infiltrated with some liquid, which is subsequently transformed into a solid with suitable mechanical properties for microtomy. The nature of the liquid–solid transformation varies. Paraffin wax, still perhaps the most widely used embedding medium, is melted and then infiltrated as a liquid which is then allowed to freeze. The plastic embedding media are infiltrated as liquid monomers, which are then polymerized to form solid macromolecules. With cryostat or frozen sectioning the matrix is the tissue's own water, frozen into ice.

Some specimens are very hard indeed, woody plant stems or bone for instance. Such materials can sometimes be cut with suitable microtome knives, if embedded in a hard medium such as a crosslinked plastic. Occasionally no embedding medium is necessary: bone may be cut directly on a sledge microtome using a tungsten carbide edged microtome knife. Other strategies are also used with hard specimens. For instance they may be softened, and then treated as routine samples: bones may be decalcified, and overfixed or overheated tissue blocks plasticized with aniline or phenol. On the other hand microtomy may be abandoned: thin layers of bone have been traditionally prepared for microscopic study by grinding and polishing.

2.3.4 Manipulating thin layers

Sometimes tissues are stained by immersion or flotation of free sections into or onto the staining reagents. More commonly such delicate tissue sections are stained, and indeed manipulated in other ways, whilst being supported mechanically, usually by a glass microscope slide. Occasionally varying sizes and shapes of other transparent materials have been substituted, for example long strips of transparent film. Problems can arise with the attachment of sections to all such supports. To enhance adhesion to glass slides a coating of gelatin or albumin is often applied (subbing). Analogously, sections of bone and of tissues embedded in plastic are sometimes attached to slides with polymeric adhesives. Unsatisfactory spreading of cells and tissue onto the slide can also cause problems. Hence such maneouvres as using spinners to give uniform blood smears, and silanizing slides to enhance spreading of metaphase chromosomes.

2.3.5 The staining process

See the rest of the book: here merely note that staining can be carried out on live or dead specimens. Live specimens may be coloured by their immersion in staining reagents, or the reagents may be ingested or injected. The more routine dead specimens may be stained before, during, or after the processes of stabilization and formation of thin layers.

2.3.6 The preservation of stained preparations

This typically involves a mounting process, sealing stained specimens into media with high refractive indices. Such a substance reduces light scattering at the specimen–medium interface, that is it reduces glare. Mounting media also isolate specimens from the environment, reducing both chemical fading and microbiological attack. Mounting media must have a low affinity for stains, or decolourization of the specimen will occur. Owing to these varied aims, mountants form a chemically diverse group of substances. As with fixatives, selection of the optimum material must be tailored to particular specimens and stains. Since tissues investigated for routine histopathology are usually coloured with water soluble reagents, they are mounted in hydrocarbon polymers such as polystyrene. This is applied as a solution which, on evaporation of the volatile solvent, forms a hard film. Specimens stained with hydrophobic reagents, such as azo dyes arising during enzyme histochemistry, would be decolourized by such media. Hence they are mounted in concentrated aqueous solutions of gelatin or sucrose. Such media, like the oil often used to mount blood smears, remain liquid.

2.4 COMPARISON OF HISTOCHEMISTRY WITH SOME RELATED TECHNIQUES

To emphasize advantages and limits of various methods of studying the content and structure of cells and tissues, histochemistry, biochemical assay and flow cytometry are briefly compared, in Table 2.2. Both the kind of information obtained and the

Table 2.2 — Some comparisons of histochemistry with related methods

	Advantages of each method (the more stars the better)		
	Histochemistry	Biochemistry	Flow cytometry
Types of information obtained			
What *is* it?	**	***	*
Where is it?	***	*	—
How *much* of it is there?	*	***	*
How *many* of them are there?	*	—	***
Problems involved in obtaining information			
Technical flexibility	***	***	*
Minimum sample size required	small	larger	small
Maximum sample size handleable	small even with automation	no upper limit	large
Minimum cost	low	moderate	high
Best type of sample	insoluble or insolubilizable	soluble or solubilizable	particulates only, e.g. cells, chromosomes, etc.

ease of obtaining it are compared. Though the table is extremely general, it does emphasize the versatility of histochemical methodologies.

2.5 STAINING THEORIES: WHY BOTHER?

There are three simple reasons why being aware and articulate about theoretical issues is a necessary skill for the practical benchworker.

Firstly the absence of theory makes the understanding of staining results difficult: "The experimental results and the control do look different, but of course the fixation schedules were not identical; do you think this could have affected the staining?".

Secondly, the absence of theory makes technical trouble shooting wildly empirical: "I agree that these Giemsa stained nuclei are blue not purple; but they often are, it might mean anything".

Thirdly, the absence of theory makes the choice of an optimum staining method, or the design of a new method, or the improvement of an old one, lack rationality: "We tried five different Methyl Green-Pyronins and none of them were up to much, so we fiddled around and came up with our own procedure".

The use of staining theories is, in fact, inescapable. When you say "You didn't use Carnoy's did you? That never works you know" this judgement comes from your own theory, of fixation in this example. Maybe imprecise, and perhaps rather unaware, but without doubt a theory.

The language of theory is important. If too remote from the concerns and experience of the practitioner then the theory will not even be considered, let alone made use of. Yet elegant theories are routinely in a foreign language; the underlying bases of histochemistry for instance are of necessity largely chemical and biochemical. Acknowledging this difficulty, this present book attempts to link such basic concepts to the practical concerns of the benchworker. Hence the organizing ideas seen in chapter headings, such as 'Why doesn't everything always stain?'.

2.6 FURTHER READING

A spectrum of articles describing the use of histochemical methods to solve problems in diverse biological and clinical fields—from botany to haematology and toxicology—are collected in *Histochemistry: the widening horizons,* edited by Stoward and Polak (1981). An excellent single-author text describing the application of staining methods to pathology is available (Underwood, 1987). Of practical manuals there is no end. I am personally reassured either by the personal touch, as found in Lillie and Fullmer (fourth edition, 1976); or by rigorously edited multi-author works, such as that of Bancroft and Stevens (1982 next edition due 1988). For the learner there are three books I find particularly useful. One, by Kiernan (1981), is aimed at the university student; another, by Vacca (1984), is directed to the trainee technician in a histopathology lab. The third is a pictorial account of some five hundred things that can go wrong when staining preparations (Thompson and Luna, 1978). Finally, the topic of tissue stabilization calls for specific mention, since many accounts are oddly oblique, and fail to focus on the practical consequences for histochemistry of unsuccessful (or successful) fixation. For attempts to achieve focus try the chapter titled 'Fixation and other loss limiting procedures' in my monograph (Horobin, 1982, Chapter 3); or the fixation chapter in Kiernan (1981). For more information look into accounts such as the reviews of David Hopwood (1985; and others cited therein), or the electron microscope oriented monograph of Hayat (1981).

2.7 TEST YOUR UNDERSTANDING

Question 1. It was stated above that histochemical staining is carried out to enable us to answer the questions: What is it? How much of it is there (or how many of them are there?)? Where is it? Moreover 'what?' questions were seen to come in several varieties: biological, chemical and so on. Use these categories to suggest the kind(s) of information sought in the following cases.

(a) By a haematologist looking at a Giemsa-stained smear of peripheral blood.
(b) By a myologist inspecting a section of skeletal muscle, stained so as to demonstrate glycogen.
(c) By a histopathologist studying a section of liver, stained with rubeanic acid to demonstrate copper ions.

Question 2. Imagine that you wish to study the distribution of glycogen and of succinate dehydrogenase in rat liver. You decide to use the standard PAS and enzyme histochemical procedures. You process samples of tissue by protocols (a) and (b); and also, owing to an oversight, protocol (c): all protocols are listed in Table 2.3. Sections resulting from all three protocols are stained by both procedures.

Table 2.3 — Question 2: Tissue processing protocols—fill in the blank columns

	Losses of		Biochemical changes in		Degradation of morphology
	glycogen	enzyme	glycogen	enzyme	
Protocol (a) fixation in neutral formalin dehydrate and clear embed in paraffin wax section stain					
Protocol (b) freeze section in a cryostat staining					
Protocol (c) immerse in neutral buffer: formaldehyde-free dehydrate and clear embed in paraffin wax stain					

Complete Table 2.3—if this is not your own copy of the book, using a copy of the table.

Question 3. How would you prepare thin layers of the following specimens?
(a) Epithelial cells of the cervix?
(b) The diaphysis of an adult human femur?
(c) A skeletal muscle and its tendons?
(d) The cells present in a pleural exudate?
(e) A human eye?

Part II

Theory of the staining process

3

Why does anything stain?

3.1 INTRODUCTION

This chapter initially considers the nature and sources of affinity, the physical-chemical basis of most histochemical staining. Other bases for staining are then briefly discussed, namely the uptake of stains due to the physiological processes of live cells and tissues, and negative staining. Some differences arising with 'catalytic staining' are clarified. As a final topic, the influences of practical dyebath variables (for instance pH, presence of salt and organic solvents) on staining are considered.

3.2 AFFINITY: WHAT'S THAT?

Let us approach this question by means of specific examples. Most histochemists say that it is the binding of labelled antibodies, or dyes, or Schiff's reagent to cells and tissues which causes staining. In these instances this is quite correct; moreover reagent-tissue binding is of wide importance in histochemistry. Sometimes, however, it is the binding of a coloured substance to itself which underlies staining, with the tissues merely initiating the process and providing an inert matrix to contain the product. Examples include pigments formed during enzyme histochemical staining, and metallic crystals deposited during silver staining. Most stain–stain binding, however, depends on the same physical-chemical factors as does stain–tissue binding.

So what is meant by affinity? In keeping with the discussion so far, some histochemists use the term as shorthand for 'those attractive forces binding a stain to the tissues'. This concept is inadequate. First because, as noted above, staining may depend on stain–stain interactions. More significantly though, staining can occur even in the absence of attractive forces involving stain or tissue. Hence we here follow another usage. This regards affinity as a measure of the tendency of stain to move from a staining solution into the tissues. From this perspective *any* process which aids such movement contributes to affinity.

It should be noted that staining can be influenced by affinity at more than one point in the process. Consider a stain such as the Periodic acid–Schiff procedure, in which colourless reagents enter the tissue and then give rise to coloured product. Both uptake of the colourless reagents into the tissue, and binding of the final coloured stain to the tissue, involves affinities.

3.3 SOURCES OF AFFINITY: STAIN–TISSUE AND STAIN–STAIN ATTRACTIONS

Such forces may be directed, for example covalent and hydrogen bonds, or non-directed, coulombic and van der Waals attractions for instance. They vary in strength and in their range of action. Van der Waals attractions are the weakest and most short range; and coulombic forces are the strongest and of longest range.

3.3.1 Coulombic forces

These are undoubtedly the most widely discusssed attractions. Coulombic forces are often significant since tissue constituents such as proteins, nucleic acids, some carbohydrates and lipids are ionic, as are many staining reagents, for example acid and basic dyes, labelled antibodies, periodate, and Schiff's reagent. Interaction of ionic reagents with tissue ions is typically discussed in terms of some form of an ion-exchange. Consider the uptake of basic dyes by cell nuclei and ribosome-rich, that is basophilic, cytoplasms. In the named sites the predominant fixed, and hence immobile, macromolecules are DNA and RNA respectively. These polymer chains carry anionic phosphate substituents. Since nucleic acids are salts, many cations are associated with each polyanion. In the living cell the cations are mostly K^+ and Mg^{2+}, whilst in fixed tissue immersed in staining baths the counterions will vary with the buffer used. Basic dyeing can thus be considered as due to the *exchange* of such small, colourless, mobile cations with the mobile coloured cations of the basic dye. Further illustrations of the role of ion-exchange in staining are shown in Fig. 3.1. Note that the example shown in Fig. 3.1(b) involves reagent–reagent coulombic attractions. These forces are involved in holding together the ionic crystals of lead sulphide. Such insoluble precipitates are not bound to the tissue, they are merely held mechanically within it.

3.3.2 Covalent bonds

These are strong and well directed. They are responsible for the binding of stain to tissue in such routine procedures as the Periodic acid–Schiff and Feulgen nucleal methods. Covalent bonds vary in character. The examples cited are typical of organic compounds, whereas those present in metal–ligand links are very much more polar. Covalencies may be reagent–tissue, as when the dye Mercury Orange is bound to tissue thiols. Alternatively a coloured product may be generated *in situ* and remain trapped in the tissue, as with the formation of a red coordination complex between tissue nickel and dimethylglyoxime. Covalencies can also be involved in reagent–reagent interactions. An example is the formation of insoluble polymeric dyes from diaminobenzidine derivatives during the enzyme histochemical demonstration of peroxidase. Examples of the role of covalent bonding are illustrated diagrammatically in Fig. 3.2.

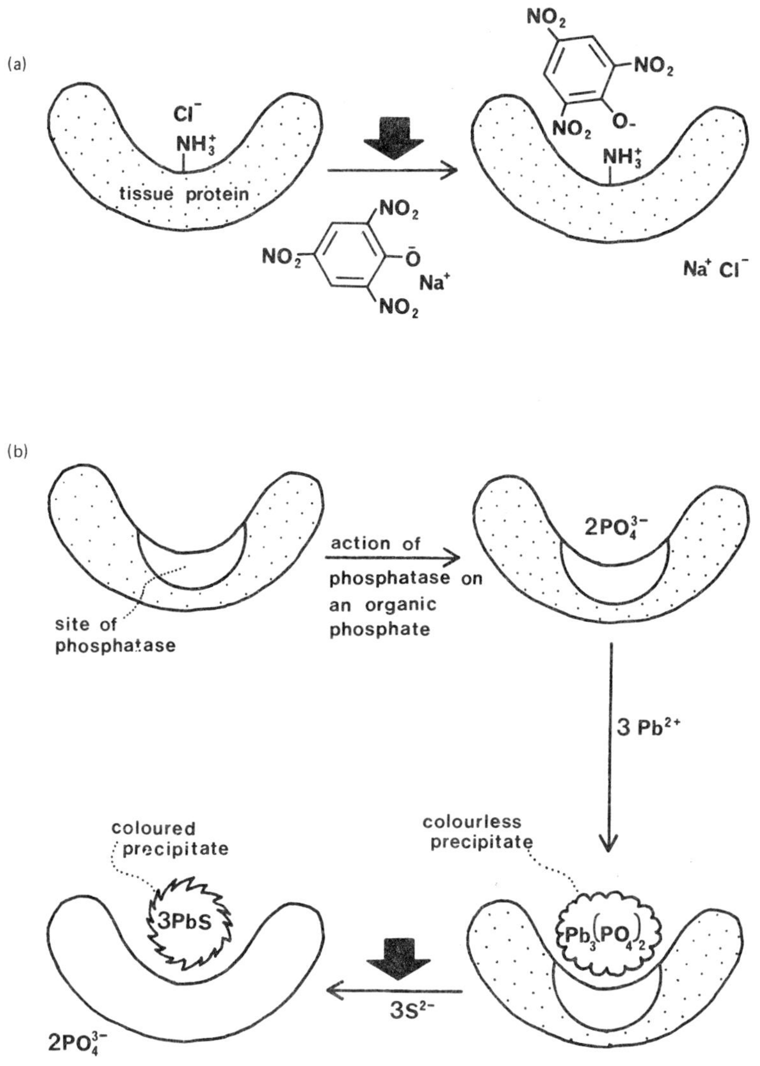

Fig. 3.1 — Examples of ion-exchange (heavy arrows) in histochemical staining. (a) In the uptake of an acid dye (picric acid) onto protein, at low pH. (b) In the visualization of the enzyme acid phosphatase by formation of a black lead sulphide deposit.

3.3.3 Van der Waals forces

Although little discussed in the histochemical literature, van der Waals forces are probably the most commonly occurring stain–tissue attractions. The weak, short-range and non-directed attractions that are collectively described as van der Waals forces include dipole–dipole, diple–induced dipole, and dispersion (London) attractions. Van der Waals forces are strongest when stain and tissue molecules possess numerous dipolar and/or polarizable chemical groupings. Some of these points are illustrated diagrammatically in Fig. 3.3.

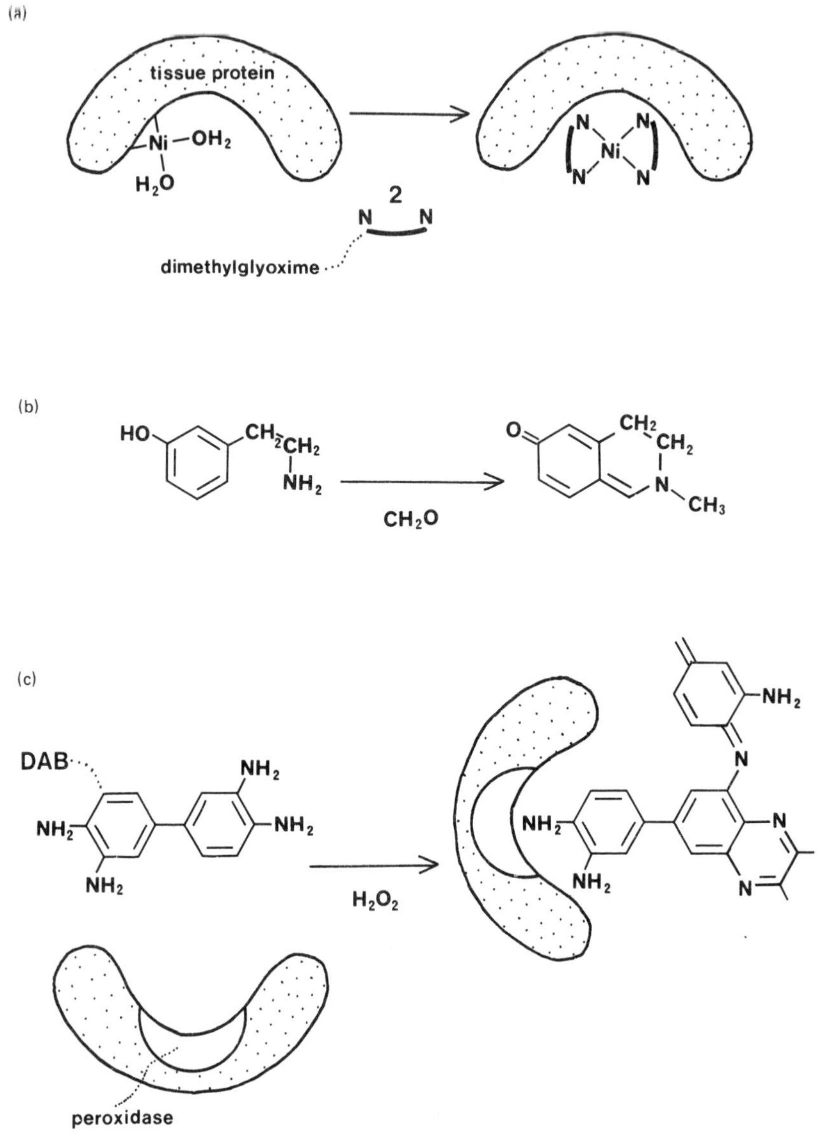

Fig. 3.2 — Staining involving formation of covalent bonds. (a) Between nickel ions, initially bound to tissue ligands, and the chelating agent dimethylglyoxime, yielding a red complex involving polar Ni-ligand covalencies. (b) Between dopamine and formaldehyde, to produce a fluorescent isoquinoline. (c) Between diaminobenzidine substrate molecules, by oxidative coupling catalysed by tissue peroxidase, giving rise to a brown polymeric dye.

Substantial polarizability and large dipoles are characteristic features of molecules with aromatic and conjugated systems; and these factors tend to become more significant as the sizes of such systems increases. Consequently a bisazo dye such as Congo Red has a higher affinity for tissue proteins than does a monoazo dye such as Ponceau 2R. The bisazo, but not the monoazo, dye will stain tissue proteins such as

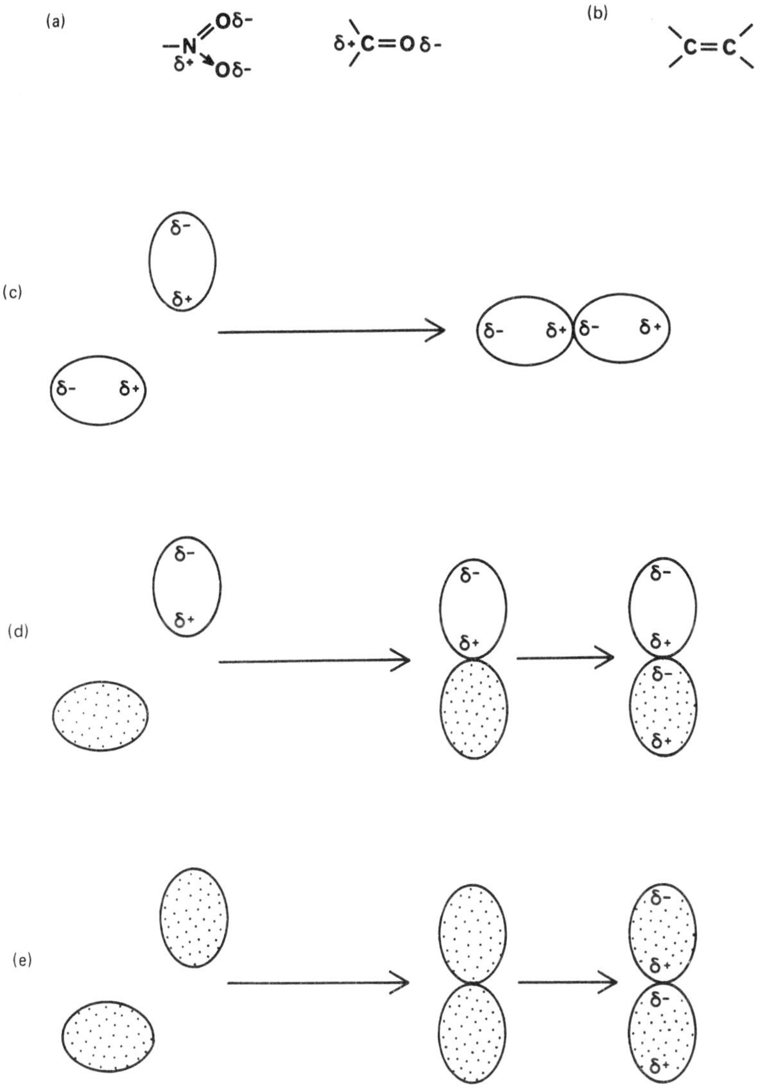

Fig. 3.3 — Van der Waals attractions: chemical fragments and contributing forces. (a) and (b) show, respectively, dipolar and polarizable groups found in many stains and biopolymers. (c), (d) and (e) show, respectively, formation of dipole–dipole, dipole–induced dipole, and dispersion (London) attractions. Note: dipoles are indicated by δ+/−, and polarizable structures by stippling.

elastic fibres even when colombic attractions are absent. Analogously, diformazans formed as the reaction products when demonstrating dehydrogenases bind strongly to tissue proteins. Consequently the diformazans give much less artifactual staining of lipids than do the equivalent monoformazans.

Differences between the sizes of the aromatic systems of these high-affinity bisazo dyes and diformazans, and the low-affinity monoazo and monoformazan compounds, can be seen by inspecting the structural formulae in Fig. 3.4. As well as visual information, this figure also gives the conjugated bond numbers (CBNs) of the four stains. CBNs are numerical structural parameters which provide another way of assessing sizes of conjugated systems, and hence of the likely contributions van der Waals forces to affinity. For a brief sketch of such parameters see Appendix 1.

Biopolymers will particularly favour van der Waals binding of stains when conjugated groupings are present. Examples are tyrosinyl and tryptophanyl amino acid residues in proteins, and the heterocyclic bases of nucleic acids. When using aqueous solutions even poorly polarizable structures, such as the alanyl or leucinyl amino acid residues, favour van der Waals forces since their lack of hydration shells permits the close contact between stain and biopolymer needed for strong attractions.

3.3.4 Hydrogen bonding

This relatively weak, but directed, form of bonding is another reagent–tissue attractive force which is occasionally suggested to explain the affinity of histochemical stains for tissues. Hydrogen bonds arise when a hydrogen atom is situated between two strongly electronegative atoms, commonly oxygen or nitrogen. Hydrogen bonds can certainly occur between stain and tissue biopolymer. However, as such bonds also occur between stain or biopolymer and water, the contribution of hydrogen bonding to affinity when using aqueous solutions is unclear. Examples of these various possibilities are shown in Fig. 3.5. If the solvent is largely non-aqueous then dye–tissue hydrogen bonding could contribute significantly to affinity; as in the case with Best's Carmine stain for glycogen. It is also possible that initial stain uptake due to, say, coulombic attractions may be followed by hydrogen bond formation.

3.4 SOURCES OF AFFINITY: A FOCUS ON ENTROPY CHANGES

3.4.1 What is entropy?

Uptake of stain into the tissues occurs because of stain–tissue affinity. Affinity, as has been discussed, can be due to a variety of reagent–tissue attractive forces. However, affinity can also be due to the *entropy* of the staining system increasing as the staining reagent moves out from the solution into the tissue.

Entropy and entropy change are complex concepts. Here it will be sufficient to regard entropy as a measure of the disorder (or untidiness) of a system, and to note that processes in which disorder increases are processes which occur spontaneously, without imputs of energy. If you remember that your laboratory bench always seems to get untidy by itself, but that you have to work at tidying it up, you have the idea of entropy-driven processes. How, though, does this relate to staining affinity?

3.4.2 Reagent entropy change

Imagine placing a cryostat section of fatty liver into an alcoholic solution of a non-ionic sudan dye. Initially there are two compartments, solution and section, with dye present in only one of them — quite a tidy arrangement. However, if the dye dissolves in the neutral lipid droplets of the lipocytes, then both compartments will contain dye, and the system will have become less ordered, more untidy. Hence the process involves an increase in entropy of the stain, and thus occurs spontaneously.

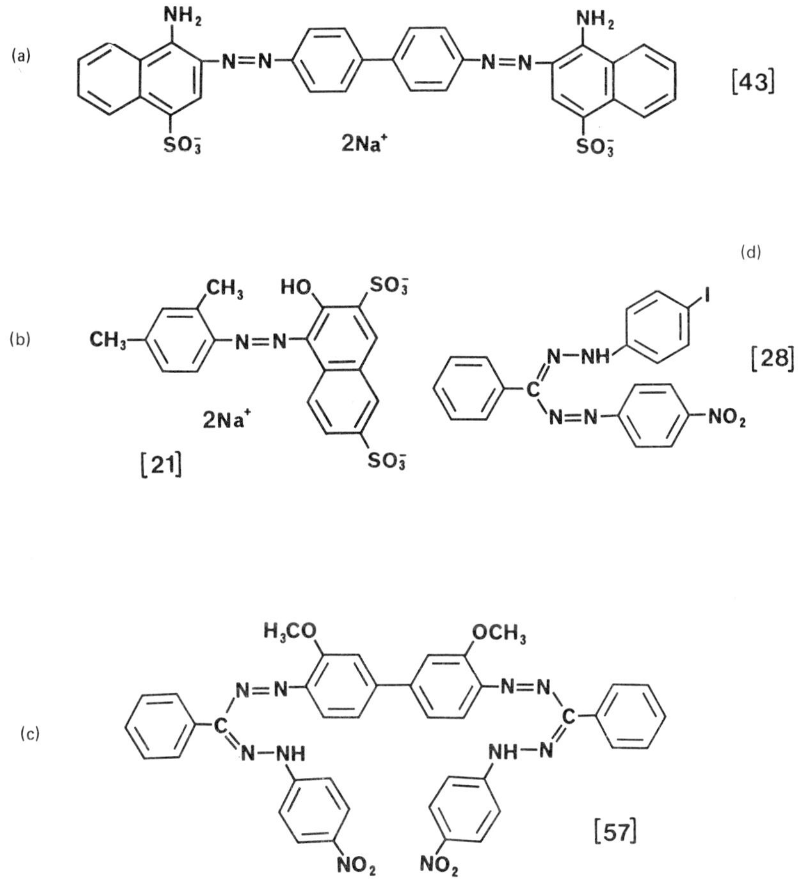

Fig. 3.4 — The size of conjugated systems: visual and numerical descriptions. (a) and (b) show, respectively, the high-affinity bisazo dye Congo Red and the low affinity monoazo dye Ponceau 2R. (c) and (d) show, respectively, the substantive diformazan derived from Nitro Blue Tetrazolium, and the non-substantive, fat-soluble monoformazan derived from Iodotetrazolium. Note: the conjugated bond numbers are shown within square brackets [so].

In this case no dye–tissue attractions need occur for dye–tissue affinity to exist. In most processes of staining, however, both entropy and attractive forces contribute to overall staining affinity.

It must also be remembered that it is the overall entropy change of the staining system which is important. Even ignoring the entropy of the stain molecules, if movement of dye into the tissue is coupled with some other disorder-generating process, then staining will occur.

3.4.3 Solvent entropy change: hydrophobic bonding

The significant example is hydrophobic bonding. This occurs only when the solvent is aqueous, since the entropy increase is due to an increase in disorder of water molecules during the staining process. Water is an odd substance, and its oddities are

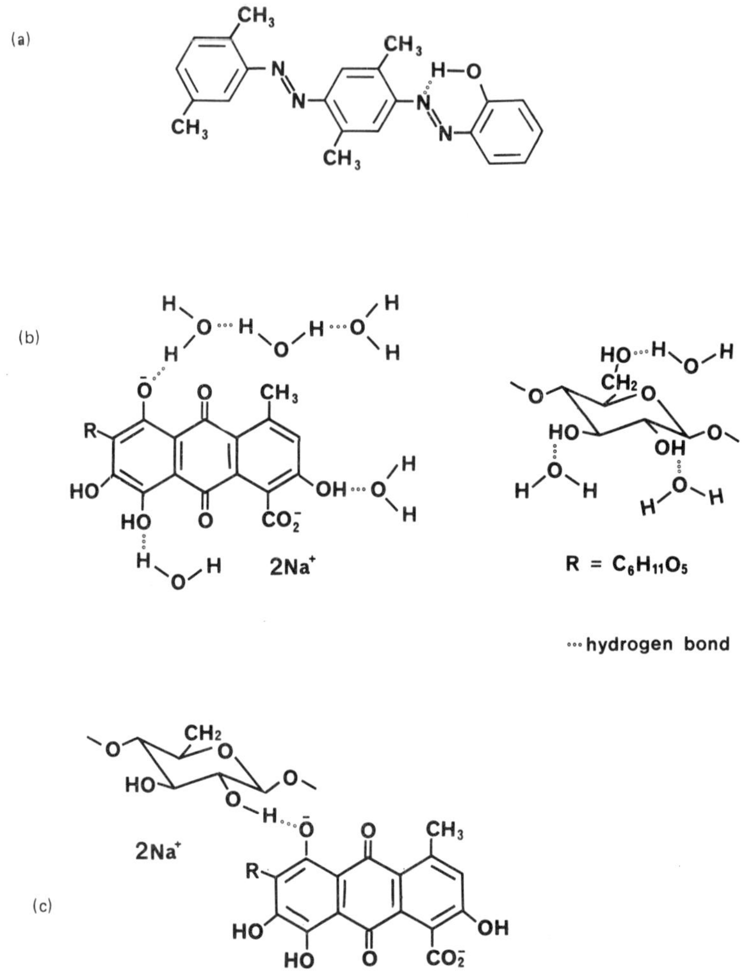

Fig. 3.5 — A variety of hydrogen bonds. (a) Intramolecular hydrogen bonding within a stain molecule, in this case the azo dye Oil Red O. (b) Intermolecular hydrogen bonding between: water molecules; water and the dye Carminic Acid; and between water and a glucose residue. This situation arises when the staining solvent is aqueous. (c) Intermolecular hydrogen bonding between Carminic Acid and a glucose residue. This situation arises, for instance, when the glucosyl polymer glycogen is exposed to a substantially non-aqueous solution of the dye. Hydrogen bonds are indicated by dotted lines.

due to a proportion of its molecules being in the form of hydrogen-bonded clusters or aggregates, as shown in Fig. 3.5(b). Such clusters are stabilized by the presence of hydrophobic solutes, such as ethyl or phenyl groupings. Thus in aqueous solutions the hydrophobic groups on dissolved staining reagents and on tissue biopolymers will be surrounded by a sheath of highly ordered clusters of water molecules. Consider removing hydrophobic groups of stain and tissue from contact with the water, and placing them in contact with each other. Some of the ordered water clusters, no longer stabilized by contact, fall apart into randomly dispersed water molecules. Consequently the solvent becomes more disordered, and there is an increase in entropy of the staining system. This entropy-driven coming together of hydrophobic groups in aqueous solution is what is termed hydrophobic bonding; it is illustrated in Fig. 3.6.

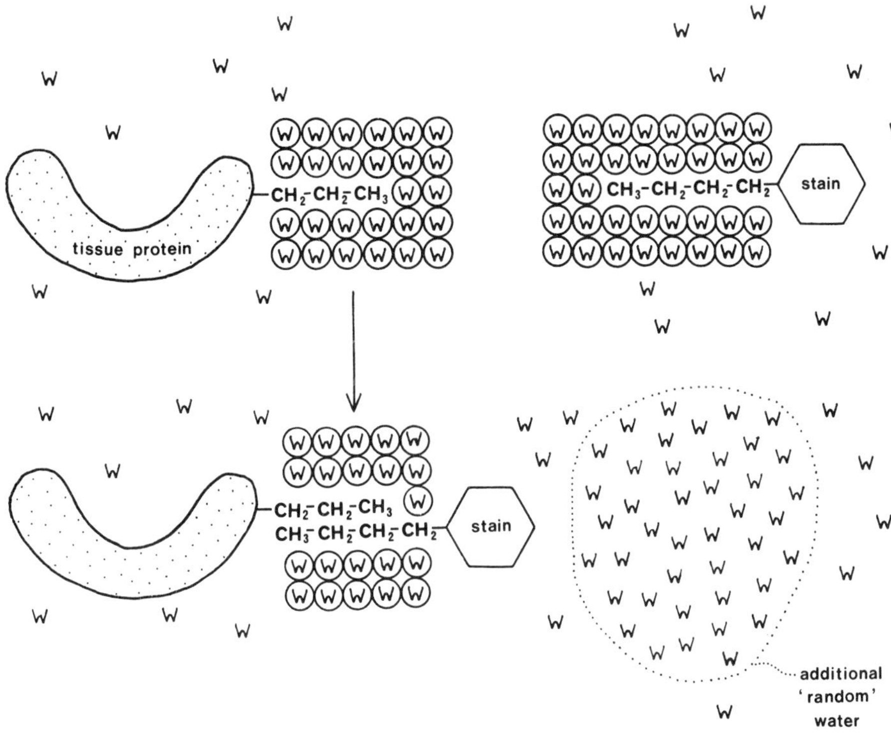

Fig. 3.6 — Hydrophobic bonding facilitating binding of a hydrophobic stain to a hydrophobic group on a tissue protein. Note: W represents monometric, randomly distributed water molecules; whereas (w) represents ice-like, ordered water.

Hydrophobic bonding will contribute to stain–tissue affinity whenever the solvent is aqueous, and the stain and the tissue biopolymers carry hydrophobic groups. As an example, consider metachromasia. This is the term used to describe the staining of different tissue substrates different colours by a single, pure, dye. An instance is staining of mast cells by Methylene Blue: nuclei are coloured orthochromatically azure blue, whilst the granules are coloured metachromatically red-purple. The orthochromatic colour is due to dye binding as single, monomeric dye ions; the metachromatic colour to clusters of dye ions stacked up on the tissue. The dye–dye binding, occurring only in certain tissue structures, adds to the staining affinity. What factors underlie the dye aggregation? Obviously not coulombic forces, which will be repulsive. Van der Waals forces will probably contribute, although since metachromasia can occur with dyes of small conjugated size this contribution may be small. However, hydrophobic bonding certainly contributes significantly to aggregation, since exposure of a stained preparation to a non-aqueous solvent, such as the dehydrating alcohol, greatly reduces metachromasia.

3.5 STAINING DUE TO NON-AFFINITY PROCESSES

3.5.1 Introduction

There are staining mechanisms in which the crucial factors are other than stain–tissue affinity. Two such mechanisms will be briefly discussed.

3.5.2 Vital staining

The physiological and biochemical processes of live cells can control uptake and distribution of stains. Classic examples of vital staining are the endocytotic uptake of dyes such as Evan's Blue, with subsequent segregation of the dye in secondary lysosomes; and the passive uptake of dyes such as Janus Green, which are then localized in mitochondria. Modern vital staining often makes use of fluorochromes. Examples include detection of mitochondria-rich neuromuscular end plates using Rhodamine 123, and tracing of neuronal pathways with Nuclear Yellow. Such reagents are more sensitive than equivalent non-fluorescent dyes, allowing low stain concentrations, and so reducing toxicity problems. Of course, whatever the implication of the term vital stains, the cellular processes themselves possess a physicochemical substratum. Perhaps for this reason such methods are nowadays often described by phases such as 'use of molecular probes'.

3.5.3 Negative staining

This is another odd-ball approach, in which the surfaces of objects are visualized by deposition of coloured material up against them. Microscopically, this enables us to visualize intra- and intercellular channels and ducts. The precipitation of Thionin picrate within the canaliculi of bone in the Schmorl's procedure is one example. Outlining unstained bacteria, fungi and spirochaetes by Nigrosin is another. The same principles are used by electron microscopists, to study such phenomena as the permeability of epithelia, and, on a molecular level, by biochemists investigating the shapes of macromolecules such as glycosaminoglycans, immunoglobulins, and nucleic acids.

3.5.4 Catalytic staining

The crucial staining steps of many of the procedures discussed in this book involve binding stain to some target in a cell or tissue. However, in some staining methods the critical property of the target is that it catalyses the formation of some visible substance. Such a reaction does involve interaction of tissue catalyst and a reagent, and the final visible reaction product must have an affinity for the tissue; nevertheless such methods do constitute a significant methodological subgroup.

Catalytic sites may already exist in live cells, or they may be induced during *post mortem* modifications. Enzymes provide examples of the former type; and demonstration of these proteins by methods exploiting their innate catalytic properties (enzyme histochemistry) is described in some detail in section 8.4. Examples of induced catalytic sites are provided by silver staining methods, as discussed in section 8.2. The essence of silver staining is the reduction of silver(I) to metallic silver(0). Here we merely note that induced catalysts of this redox reaction include crystals of metallic sulphides, formed from low concentrations of metal ions present in live cells, and sub-microscopic aggregates of silver atoms formed by prior action of tissue reducing agents on silver(I) cations. Moreover some stains can themselves act as catalysts of the reduction process; hence silver staining can be used to amplify a prior staining of some other kind. An example in which the initial staining procedure is immunostaining is discussed in section 8.5. It is interesting that another traditional method, Romanowsky–Giemsa staining, also involves a catalytic step (see section 8.1.1).

3.6 INFLUENCES OF PRACTICAL STAINING VARIABLES

Remember that this chapter is mostly concerned with reagent–tissue affinities. Consequently staining time, which influences selectively but not affinity, will not be discussed here. Practical variables such as mode of fixation and the pH of the staining solutions influence both affinity and selectivity, so they get double billing. Some influences of pre-staining and post-staining treatments on affinity are discussed in Chapter 6.

One way of altering the stain–tissue affinity is to alter the chemical structure of the stain. Examples of possible structural changes include electric charge (like charges attract), the size of the conjugated system (the larger this is, the stronger will be van der Waals attractions), and the hydrophilic/hydrophobic character (hydrophobic bonding is favoured by hydrophobic reagents). Naturally, affinity due to hydrogen bonding or covalent bonds will require appropriate groupings on the stains.

The other inevitable component of the staining bath is the solvent, which may be aqueous or non-aqueous. Water is a necessity for hydrophobic bonding to occur; on the other hand water will compete with stain–tissue hydrogen bonds.

Staining solutions may contain co-solutes which influence affinity. Acids and bases can alter coulombic contributions to affinity by changing the electric charges on the tissues and stains. For instance most proteins carry an overall negative charge at high pH, but an overall positive charge at low pH. Neutral salts also influence coulombic contributions to affinity, in an extremely complex manner. Ions such as Al^{3+} and Cr^{3+} can bind to tissue biopolymers and hence alter their electric charge. Even if ion-binding does not occur, high salt concentrations suppress the surface electrical potential on tissue polyelectrolytes, which can result in both decreases ('competition') or increases ('screening') in staining. Finally note that salts can cause aggregation of dyestuffs; and such large particles may fail to stain, or stain with reluctance.

A variety of other co-solutes are sometimes found in staining solutions, with a corresponding variety of influences on affinity. For instance hydrophobic bonding is reduced by the presence of aniline, phenol or urea; detergents may compete with stains for sites on the tissue, or may form non-staining complexes with the stain in solution.

3.7 FURTHER READING

Many of the histochemical texts mentioned in section 2.6 (for instance Kiernan, 1981) discuss, if rather briefly, such affinity contributions as coulombic forces and hydrogen bonding. However, such accounts are typically spread amongst many technical methodologies, such as immunostaining or bacteriology. An integrated discussion of affinity, and of the applicability of the concept to histochemical staining reagents in general, is not easy to find. For an extended account of affinity, and its applicability to such diverse procedures as metal impregnation and enzyme histochemistry as well as to the uptake of dyestuffs, see my *Histochemistry* (Horobin, 1982), which text also discusses vital staining. An admirable sourcebook for the chemical structures and properties of dyes and certain other histochemical reagents is *Conn's biological stains* (Lillie, 1977).

3.8 TEST YOUR UNDERSTANDING

Here you can practice assessing which properties of reagent, staining solution, and tissue influence affinity. A few tips may be useful. In each case, assume that staining reaches equilibrium, and so is affinity controlled. Start by establishing the structures of the stains, which in this case are all dyes. Then decide, for each group of dyes, the relative values of the various properties which could influence affinity. Consider electric charge, size of the conjugated system, hydrophilic/hydrophobic character, and the presence/absence of molecular groupings capable of forming covalent or hydrogen bonds.

Question 1. Imagine you have immersed tissue sections in two dyebaths, both at the same acidic pH. One bath contained Alizarin Red S, the other Trypan Blue; both dyes were at the same molar concentration. Then you stained more sections with these two dyes, this time from alkaline solutions.

At low pH, both dyes gave an intense overall colouration of all protein-rich structures. At high pH, Alizarin Red S failed to stain, whereas Trypan Blue gave rise to a weak colouration. To what affinity effects can you attribute:

(a) The fall in staining intensity seen with both dyes, as the pH rose?
(b) The difference between the two dyes?

Question 2. Again, imagine you immersed tissue sections in two acidic aqueous dyebaths. One solution contained Methylene Blue, the other Victoria Blue B. Then you stained more sections with the same concentrations of these two dyes; but this time the solvent used was acid alcohol.

The completely aqueous solutions gave strong colouration of nuclei and other basophilic components. The alcoholic solutions gave staining of reduced intensity, with that of Victoria Blue being grossly reduced. To what affinity effects can you attribute:

(a) The fall in staining, for both dyes, as alcohol was added to the dyebath?
(b) The difference between the two dyes?

Question 3. This time, imagine that you wanted to demonstrate the enzyme succinate dehydrogenase within a cryostat section of skeletal muscle. For a visualizing agent you proposed using a tetrazolium salt. As a technical preliminary you investigated the three different tetrazolium salts Benzothiazolylphthalhydrazidylstyryl tetrazolium (BPST), Iodonitro Tetrazolium (INT), and Nitroblue Tetrazolium (NBT). Other practical factors were kept constant.

The staining outcomes varied markedly. The formazan derived from NBT was deposited as discrete granules within muscle fibres. The formazan derived from BPST also produced intracellular granules, although a substantial amount of coloured product was seen to diffuse away from the section into the incubation medium. The formazan from INT was deposited intracellularly as large crystalline masses, and also produced colouration of neighbouring fat cells.

Assuming that these differences reflect only variations in the tissue–formazan affinities, what explanations can you offer?

4

Why doesn't everything always stain?

4.1 SENSITIVITY, SELECTIVITY AND OTHER USEFUL CONCEPTS

Selectivity is a one-word version of the chapter heading, but why sensitivity? Primarily because, in practice if not in theory, differences in sensitivity are often hard to distinguish from differences in selectivity.

Consider a simple case: staining a virally infected cell with a sensitive cationic fluorochrome, producing intense fluorescence of nuclear chromatin, and weak fluorescence of clumps of virus particles. Viewing the same infected cell after staining with a less sensitive reagent, such as a non-fluorescent basic dye, you might only have seen staining of nuclear chromatin. To the eye the less sensitive stain seems the more selective, yet delicate instrumentation would show the same staining pattern with both reagents. The difference is due to sensitivity, not physicochemical selectivity.

A further blurring of the boundaries between sensitivity, selectivity and your perceptions can arise from the distribution of stain within cells and tissues. Consider two cells containing equal amounts of some enzyme, which you demonstrate histochemically. In one cell the enzyme occurs within organelles large enough to resolve in the microscope. On inspecting this you night say 'Ah look, granules'. In the other cell the enzyme is evenly distributed within the cytosol. You may now comment, 'Oh, just background staining'.

Selectivity dùe to physicochemical effects can itself derive from several quite different sources. These are briefly outlined below; some are subsequently discussed in more detail.

Staining may or may not involve localized sites. Techniques such as the periodic acid–Schiff procedure, or fluorescent-labelled antibody methods, can be regarded as producing stain bound to localized molecular sites. Staining neutral lipids by dissolving sudan dyes in fat droplets cannot be regarded in this way. Enzyme histochemistry does involve distinct catalytic sites; however, the stain generated by an enzyme is not bound to the catalytic molecule. Other things being equal, the larger the number of binding or catalytic sites (or the greater the mass of tissue substrate) within a given structure, the stronger is the staining. Staining of blood smears with Romanowsky dyes provides an example. Intense red staining, due to binding of Eosin by proteins, is seen only in dense eosinophil granules and erythrocytes; the more dispersed proteins occurring throughout the cell cyoplasms are not coloured bright red.

Next consider two structures having equal numbers of binding sites: for simplicity of expression, the site/non-site distinction is ignored in this and the following paragraphs. If the sites in one structure have a high affinity for a stain, and the sites in the other structure a low affinity, and if staining is allowed to approach equilibrium, then differential staining of the first structure occurs. Reactivity or enzymic activity are equivalent to affinity in this case.

Then there is the case of two structures possessing equal numbers of binding sites for a stain, with the sites having equal affinities. Selective staining can still occur if the rate of penetration of a staining reagent into the two structures differs; or if the rate of reaction of a reagent differs in the two structures; or if the rate of loss of stain from the two structures differs. In everyday laboratory practice staining usually does not have time to reach equilibrium, and differentiation is routine, so such rate effects are commonplace.

Finally there is the possibility that the number and affinities of sites are the same in two structures, and that there are no rate effects. Differential colouration can still occur, owing to interactions of bound stain molecules with themselves or with the site. Examples include the selective metachromasia of dyes, and the quenching of fluorochromes.

So physicochemical selectivity may depend upon differences in the numbers of binding sites, affinities, rates, or stain–stain and stain–site interactions. The selectivity of real world methods, as will be seen in Chapter 8, are routinely influenced by more than one factor. For clarity the following discussion starts with a simple case.

4.2 AFFINITY AND SELECTIVITY

4.2.1 The site model and its limitations

As already noted, for two structures with equal numbers of binding sites, the occurrence of different stain–site affinities in the two structures may result in affinity-controlled selectivity. An example is selective glycogen staining using Best's carmine. This stain employs a largely non-aqueous solvent, encouraging hydrogen bonding between the dye and hydroxyl groups on the polysaccharide. This phenomenon explains the staining of glycogen, but does not explain the lack of staining of adjacent cytoplasmic proteins, which also carry numerous hydrogen bonding groups. Doubtless a partial explanation of the selectivity is that hydroxyl groups typically form stronger hydrogen bonds than do amino and sulphydryl groups; and whilst nearly all hydrogen bonding sites in glycogen are hydroxyl groups, this is not so in proteins.

Reasonable though this sounds, a limitation of the site model must be acknowledged. The model is most useful for understanding affinity-controlled selectivity in those cases where site–stain interaction is intense, for example with site–stain covalencies. Hydrogen bonds however are weak. To be confident of understanding the Best's carmine procedure we should ask if other differential affinity factors act to reinforce the staining pattern occurring due to hydrogen bonding?

A plausible affirmative answer is found when the role of coulombic forces are considered. Under the mildly alkaline staining conditions used, glycogen will be non-ionic. Polypeptide chains of proteins will carry overall negative charges, due to the carboxylate anions of acidic amino acid residues. The dye, Carminic Acid, is also anionic. Although coulombic forces will not assist dye–glycogen interactions, they will markedly reduce dye–protein affinity and so increase staining selectivity.

4.2.2 Selective affinity: generalizations and examples

Except for staining involving formation of covalent bonds, and for certain labelled antibody methods, simple site models are difficult to apply. To understand affinity-controlled selectivity we more typically need to ask: what are the various contributions to stain–tissue affinity, and do these act differentially in stained and non-stained structures? Moreover we need to remember that affinity contributions can involve tissue, stain, solvent, and co-solutes. To illustrate these points two cases will be examined.

The first is the selective staining of neutral lipids by non-ionic dyes, such as the sudans. Assume the use of cryostat sections, say of fatty connective tissue. Assume that a hydrophobic dye, such as Oil Red or Sudan Black, is applied from an isopropanol solution. To simplify the discussion, assume that there are no substantial dye–solvent, dye–lipid or dye–dye interactions. In this simplified system staining of lipids is readily accounted for by the increase in entropy resulting from distribution of dye in two compartments (that is solvent and lipid) rather than merely in one. As the same would be true for staining of non-fatty components of the connective tissue, it is necessary to enquire as to the difference between fat and non-fat. One obvious difference is that fat is virtually water-free, whereas fibres and cells are largely composed of hydrated bipolymers. Putting a hydrophobic dye into such hydrated environments would be entropically, and possibly enthalpically, very unfavourable. Consequently this does not occur, thus giving rise to the useful staining selectivity.

As a second example consider the localization of final reaction products generated when demonstrating a mitochondrial enzyme, such as succinate dehydrogenase, by a tetrazolium-formazan method. The first visualizing agents of this type used histochemically were monoformazans such as Iodotetrazolium. These gave rise to formazan pigments with poor localizations. Whole cells were stained, not individual mitochondria; moreover non-enzymic staining of fat cells often occurred. Subsequently bis-tetrazolium salts (for instance Nitro Blue Tetrazolium) were adopted, since their formazans could visualize single mitochondria, and gave little or no artificial lipid staining. Both types of formazan are hydrophobic, so the increased selectivity was not due to the effects discussed in the first example. However, histochemically applied bis-formazans have conjugated systems about twice as large as mono-formazans: see Fig. 3.4. Hence the high selectivity of bis-compounds can be attributed to stronger van der Waals forces between their large conjugated systems and the tissue proteins surrounding enzymic sites. There is a further strategy for controlling artifactual fat staining, namely to use a tetrazolium salt giving rise to a hydrophilic formazan, such as BPST. This lacks lipid solubility, and does not stain fat cells. However, its conjugated system is not large, so BPST is not tightly bound by van der Waals forces to the tissue proteins; indeed it sometimes diffuses out of the tissue section into the incubation medium.

4.3 DIFFUSION AND SELECTIVITY

4.3.1 Principles, causes and general rules

In many cases the rate of reaction of a stain with tissue is extremely fast compared to the rate of diffusion of the staining reagent through the tissues: exceptions are discussed in section 4.4. Consequently selective staining patterns are often controlled by diffusion rate, rather than by affinities. Sometimes of course selectivity is influenced by both affinities and diffusion rates.

As a simple example consider two tissue structures, A and B, with equal numbers of binding sites, which sites have equal affinities for a given staining reagent. Assume this reagent diffuses slowly through structure A, but rapidly through structure B. When the time of staining is sufficiently extended for equilibrium to be reached, the system will be affinity controlled. In the present example affinity control would result in the non-selective staining of both A and B. If, however, the period of staining is short compared to the equilibrium staining time, then a basis for diffusion controlled selectivity exists. Therefore *progressive staining* could be used. Sections are removed from the staining reagent after the reagent has diffused into structure B, but before any significant penetration into structure A has occurred. Alternatively a strategy of regressive staining could be adopted. In this case sections are initially overstained, giving non-selective colouration. *Differentiation*, that is controlled destaining, into some suitable solvent, is then carried out. De-staining is terminated after stain has diffused out of structure B, but whilst structure A still remains coloured. Diffusion control can therefore be used to selectively stain either structure A *or* structure B.

The reason for different permeabilities of structures A and B must now be discussed. Diffusion control is control by the slowest step, or steps, in the journey of the reagent from staining solution to the final stained tissue structure. Thus the rate of diffusion across the solvent–tissue boundary, or the rate of diffusion through the solid tissue, or (in plastic sections) the rate of diffusion through the embedding medium can be the slowest step. Diffusion controlled staining thus depends on factors ranging from tissue morphology to the physicochemistry of the constitutent biopolymers. Examples of the more important influences will now be discussed.

Starting with the most visible factor, the greater the surface area of a given volume of specimen the faster the transfer of stain from solution into the bulk of the tissue. Variations in surface area arise in a number of ways, for example one technical factor markedly influencing surface area is fixation; as is graphically shown in Fig. 2.2. After ispecting this figure, it will not surprise you that tissue fixed in Carnoy's fluid stains faster than similar tissue fixed in neutral formalin. Other factors are discussed at length in Chapter 7.

At one remove from the microscopically visible, consider the influence of tissue density on diffusion rates of staining reagents. Other things being equal, the higher the density the slower the rate of reagent diffusion. In cryostat sections, or in dewaxed sections taken down to water, the denser structures include erythrocytes, zymogen granules and regions of calcification; and structures with much lower densities include cartilage matrix and goblet cell mucin. Such densities are not immutable, but are influenced by technical factors such as fixation, embedding, and the conditions of staining. Thus aqueous stains typically cause swelling of collagen fibres, especially at extremes of pH; whilst elastic fibres remain virtually unswollen. Such swelling is fixative-dependent. Chromosome spreads fixed with methanol–acetic acid can become enormously swollen; whilst formalin or glutaraldehyde fixation dramatically reduces their plasticity. The elevated temperatures occurring during paraffin embedding can increase the density of tissue components such as erythrocytes, an effect not occurring in tissue embedded at lower temperatures, for instance in a plastic medium such as glycolmethacrylate. However, such plastic media, unlike paraffin, are not removed prior to staining. Hence the synthetic polymer itself increases the overall density of a section cut from a plastic embedded tissue block.

Finally note some molecular features of biological structures which influence diffusion rate, often via influences on density. It is the polysaccharide-rich structures, especially polyanions such as heparin and chrondroitin sulphate, which are most hydrated and swollen in water, and so are easily penetrated by staining reagents. The swelling of tissues at extremes of pH is largely a function of the number of ionized groups carried by their protein components. Thus collagen, a typical protein, swells grossly at both high and low pHs; an effect less marked with elastin, an unusual protein having few ionizable amino acid residues. Of course swelling can also be restricted by the presence of crosslinks between biopolymers, whether native or fixative-induced. Consequently a protein such as keratin, which contains numerous disulphide bridges, does not swell appreciably in aqueous solvents and is slow staining.

Thus far attention has been directed at properties of the tissue. However, the chracteristics of the staining reagents, and of reagent–tissue interactions, are also influential. The simplest feature of reagents which affects rates of diffusion is size. Diffusion rates fall markedly as sizes of reagents increase. Qualitative impressions of reagent size can be obtained by eyeballing structural formulae. However, structure parameters such as molecular or ionic weights are more easily referred to; look at the acid dyes in Fig. 3.4, and note that the anionic weights of Ponceau 2R and Congo Red are 434 and 651 respectively. High stain–tissue affinities also reduce diffusion rates, since a reagent tightly bound to tissues is not free to move. An example already discussed is the difference in localization between the mono- and bis-formazans generated in enzyme histochemical stains.

After considering these multitudinous details, it is possible to generalize about which cellular and tissue structures are expected to stain rapidly, and which slowly, in diffusion controlled methods. Apart from the reagent structure there are three major influences on diffusion rates during staining: the tissue itself, the aqueous or non-aqueous nature of the solvent, and the presence (or absence) of embedding medium. This last topic is discussed in Chapter 6.

'Naked' preparations, in which no embedding medium is present during staining, form a distinct category. Examples include cryostat and dewaxed paraffin sections. Here differences in density and swelling of the tissue components dominate. Aqueous staining systems form a distinct sub-category, since water causes marked swelling of structures rich in glycoproteins, glycosaminoglycans, and polysaccharides. Consequently typical relative rates of staining of tissue substrates in naked tissues, using aqueous solutions of staining reagents, are:

Fast-staining structures — cartilage matrix, and glycosaminoglycan-rich secretory granules such as goblet cell mucins and many leucocyte lysosomes; collagen fibres, and other glycoprotein structures such as thyroid colloid.
Intermediate — chromatin, muscle cytoplasms.
Slow-staining structures — Elastic fibres, erythrocytes, and other dense structures such as myelin sheaths, ribosomes, and zymogen granules; also crosslinked proteins such as the keratin in hairs and certain epithelial cells.

Staining systems using solvents which limit or prevent tissue hydration (e.g. acetone, ethanol, and methanol) differentially reduce swelling; and therefore reduce the staining rate of the fast-staining structures listed above. This has various practical consequences. For example when alcohol is used to differentiate stained central nervous system preparations, unswollen collagen fibres can lose stain as slowly as do myelin sheaths.

Reagent diffusion into — or out from — the section will of course decrease as the size of the reagent increases.

A final note: even when diffusion control is clearcut, the image perceived is only indirectly related to the mechanism giving rise to it. As an observer, you look down through specimens which are usually in the form of thin slabs. Hence a slow-diffusing, high-affinity reagent could give rise to the same picture as a fast-diffusing, low-affinity stain.

4.3.2 One stain, or several?

Rate-controlled selective staining may involve a single staining reagent, or several. If more than one, they may be applied simultaneously as a mixture, or individually in a sequence. Rate-controlled selectivity may be achieved by intent ('differentiate under microscopic control'), or unwittingly ('rinse quickly in water'). Rate-controlled selectivity may involve progressive or regressive staining. To illustrate such possibilities, several examples will be discussed.

An instance of diffusion rate control involving a single, progressively applied reagent is the selective staining of goblet cell mucin by Alcian Blue; and also the less studied Astra Blue, colloidal iron, and High Iron Diamine stains. These cationic reagents stain mucins without concomitant colouration of cell nuclei, unlike the outcomes after staining with basic dyes such as Basic Fuchsin or Toluidine Blue. Such selectivity is due to the mucin stains being large and hence slow-diffusing species, which slow diffusion maximizes the differences in staining rates between mucins and nuclear chromatin. Thus only the faster staining tissue components (such as mucins) have time to be stained. In keeping with this, selectivity is lost if staining times are over-extended.

An example of diffusion rate controlled *de*-staining (that is differentiation), involving a single staining species, is the classic iron Hematoxylin procedure for demonstrating striations in cardiac and skeletal muscle. In this case the initial uptake of stain is non-selective. Differentiation in dye-free solvent then removes stain most quickly from the most permeable zones, namely the I bands; leaving stain in the less permeable A bands and Z lines. In keeping with this mechanism, stain is also retained in other poorly permeable entities, such as erythrocytes, nucleoli, and thicker regions of the section produced by creasing and folding. It is intriguing to note how often the role of such de-staining is either overlooked or not explicitly acknowledged. Thus step 1 of many multi-step methods (such as the Baker acid–Hematein stain for choline-containing phospholipids, or the Naphthol-Pyronin visualization of horseradish peroxidase labelled antibodies) often gives general tissue staining. Selectivity is largely dependent upon subsequent differentiation.

Another classic method, namely Trichrome staining, provides us with examples of progressive staining with mixtures of dyes, whose selectivity is controlled by diffusion-rates. Trichromes colour collagen fibres and smooth muscle cytoplasm in contrasting colours; even though both of the dyes making up the staining mixtures have an affinity for both tissue components. Selectivity is dependent on two factors: (1) that collagen fibres are more permeable than are muscle cytoplasm; and (2) that the two acid dyes present in the one-bath trichromes are of very different ionic weights. In the limited staining times needed in practical methods, the larger, slower diffusing dye (for example Sirius Red, anionic weight 1235) only has time to diffuse into the collagen fibres. Although the smaller, faster-diffusing, dye (for example picric acid, anionic weight 228) has time to penetrate collagen fibres and muscle cytoplasm, it is not uniformly distributed in the final preparation, for reasons of both affinity and rate. In the collagen fibres the small dye will be competing with the larger one. Moreover even those small dye ions which are bound will diffuse out of the collagen fibres faster than out of the cytoplasm during the post-staining rinse. For picro-trichromes such differential losses of picric acid also occur during alcohol dehydration. In keeping with such a mechanism, over-extending staining times reduce selectivity; and over washing, or lengthy alcohol dehydration, can remove some or all of the smaller (plasma) stain.

So far in this section we have discussed examples involving single stains, or two stains applied simultaneously as a mixture. However sequence staining is also widely used, and may also be controlled by diffusion rates. Indeed many trichromes involve applying dyes as sequences instead of, or as well as, mixtures. In these cases the crucial factor is the ease of diffusion through the tissue components. Although these methods do typically employ large dyes, this is probably to extend staining times to technically convenient lengths. In a method such as Lillie's Biebrich Scarlet–Methyl Blue trichrome, the initial staining with Biebrich Scarlet colours both collagen and muscle cytoplasm red. Subsequent treatment with Methyl Blue is then continued only until the red dye is replaced in collagen by the blue dye. If the second stage is continued too long, all structures become blue.

4.4 REACTION RATES AND SELECTIVITY

For the sake of simplicity we will consider a case in which a staining reagent has equal affinity for two tissue structures, and diffuses into them at equal rates. Once having arrived in the tissue, however, the reagent may react with structure one much more rapidly than with structure two; so giving staining whose selectivity is based on *reaction* rates.

Examples are provided by enzyme histochemistry. Imagine a cryostat section incubated in a aqueous solution of an organic phosphate, at around pH 7. The ester may be slowly hydrolysed by both acid and alkaline phosphatases; so giving rise to weak staining of both lysosomes and brush border for instance. Selectivity may be achieved by exploiting the pH dependence of these two enzymes. Thus if a low pH incubation medium is used, acid phosphatase is more effective; the rate of hydrolysis of ester increases in the lysosomes, and decreases in the brush border. Use of high pH reverses this pattern of selectivity.

The role of reaction rates in controlling selectivity may be more subtle. Thus with certain silver stains the initial treatment of a tissue section with silver results in the non-selective binding of silver cations to tissue proteins, plus a selective formation of sub-microscopic nuclei of metallic silver at certain sites. In a subsequent treatment with a reducing agent, the silver nuclei act catalytically to transform much of the protein-bound Ag^+ into optically visible deposits of metallic silver. This requires the diffusion of Ag^+ through the tissue to the catalytic sites. However, if the reducing agent is too rigorously reactive, then Ag^+ is converted into metallic silver non-catalytically, giving rise to generalized non-selective staining. For more details of such silver staining methods see section 8.2.

4.5 METACHROMASIA, POLYCHROMASIA AND QUENCHING OF FLUORESCENCE

A staining reagent may bind in equal amounts to several cellular or tissue components. Nevertheless the colours of the stain bound at the different sites may vary, giving selective staining.

Metachromasia is a phenomenon in which a single dye or fluorochrome gives rise to two or more different colours when bound to different tissue components. Most commonly metachromasia involves absorption of light, as with the use of Toluidine Blue to stain mast cell granules metachromatically purple, contrasting with the blue, orthochromatic, staining of cell nuclei. However, metachromatic fluorescence also occurs. Thus Acridine Orange can give rise to green nuclear fluorescence, contrasting with red emission from dye bound to RNA-rich cytoplasms. In either case the underlying mechanisms involve the tissue-mediated formation of dye–dye aggregates in the metachromatic sites.

The term metachromasia thus applies to multiple staining colours arising from the application of a single dye or fluorochrome, whose binding does not involve formation of covalent bonds or changes in oxidation state. In the present account, all other multiple colouration following the application of single stains is termed *polychromasia*. The mechanisms underlying polychromasia are very varied, and here three examples will be discussed. First the selective staining of metal ions using potassium ferricyanide. This reagent yields coloured precipitates with various metal cations; ferrous iron yields a blue derivative, whereas the lead salt is red. Second the demonstration of reticulin by a silver method, such as that of Gordon and Sweet. Both reticulin and other collagen fibre types yield final reaction products comprising microcrystals of metallic silver. However, the size of these crystals, and thus their colour, varies, resulting in black reticulin and brown collagen fibres. Finally the demonstration of thiol groups using a reactive dye, aminobromanthraquinone sulphonic acid. This reagent is not chemically specific for thiols, and also reacts with amino and hydroxyl groups. However, selective staining of thiols is obtained since their reaction products are orange, whereas those resulting from amines and alcohols are greenish blue.

A conceptually related process occurs when a reagent binds at several tissue sites but only fluorescences at one. In this instance selectivity depends on the selective *quenching of fluorescence.* An example is the Q-banding of chromosomes. In this procedure the staining of isolated chromosomes with Quinacrine mustard, or a related dye, produces chromosome-specific patterns of bands interspersed with non-fluorescent interbands. Contrary to the beliefs of the inventors of such methods, this staining pattern does not derive from selectivity in dye binding, which occurs quite uniformly along the chromosomes. Instead the staining pattern reflects differential quenching of fluorescence.

4.6 INFLUENCES OF PRACTICAL STAINING VARIABLES

One way to influence selectivity of staining is to alter the structure of the reagent(s) used, a topic already discussed in section 3.6. However, a generalization pertinent to selectivity is that hydrophobic reagents with large conjugated systems, applied from largely aqueous solutions, will tend to overstain, 'to give excessive background staining'. Hydrophilic reagents with small conjugated systems on the other hand will tend to be specific, or to understain. This effect may be expected whatever the physicochemical basis of the specific staining.

Reagent structure also influences the rates of diffusion, as these tend to be inversely proportional to ionic or molecular size. To obtain selective, diffusion-rate controlled staining patterns the reagent is typically of large size.

Solvent and co-solute can also influence diffusion rates, for instance by inducing swelling or shrinkage of the tissue, and/or aggregation or dis-aggregation of the reagent. Thus dimethylsulphoxide will swell tissues and dis-aggregate dyes; as will aniline and phenol; whereas ethanol dis-aggregates dyes but shrinks tissues. Organic solvents and co-solutes can also influence selectivity by reducing or eliminating the contribution of hydrophobic bonding to affinity. So if stain–target affinity is not dependent upon hydrophobic bonding, whereas background staining is, addition of water-miscible organic solvents to aqueous staining solutions may increase selectivity. Examples include a number of classic elastic fibre stains (such as Aldehyde Fuchsin, Orcein, and Resorcin-fuchsin) which are applied from ethanol–water solutions.

Diffusion rate is also increased by raising the temperature, as in the microwaving procedures. Prestaining and post staining procedures can influence selectivity of staining, as discussed in Chapter 6.

4.7 FURTHER READING

The selectivity of histochemical staining is discussed as a general phenomenon in my monograph (Horobin, 1982, especially sections 4.7–4.20; but see also the index entry 'selectivity').

The selectivities of many of the methods used as examples in the discussion above — e.g. formazan methods in dehydrogenase histochemistry, immunostaining, Trichrome stains — are discussed in later chapters. Of course many are also discussed in the general texts cited previously (Bancroft and Stevens, 1982; Kiernan, 1981; Lillie and Fulmer, 1976; Vacca, 1984).

Further information concerning the differential densities and permeabilities of various tissue components can be found in Baker's monograph (1958, Chapter 12) and Goldstein's papers (1963, 1965). These, together with information concerning the effects of solvents and co-solutes on tissue permeability, are summarized in Chapter 4 of my monograph (Horobin, 1982).

4.8 TEST YOUR UNDERSTANDING

Question 1. Cells grown in culture were homogenized, and the homogenate fractionated by centrifugation. Biochemical assay of the fractions indicated that the enzyme Z was present in the cytosol. Various histochemical investigations concerning the distributions of Z were carried out on intact cells still mounted on coverslips. These studies used a commercially available monoclonal anti Z immunoglobin (anti Z IgG). The results are shown in Table 4.1. Suggest explanations for the various discrepancies. Which results would you trust, the histochemical staining or the biochemical assay?

Table 4.1 — Immunostaining of cultured cells by a variety of regimes

Treatment of cells	Staining appearance	Morphological appearance
Regime 1:	No staining	Poor
(a) Stain with anti Z IgG		
Regime 2:		
(a) Expose to pronase (a protease) for 1 min	Cytosol stained	Poor
(b) Stain with anit Z IgG		
Regime 3:		
(a) Expose to pronase for 1 hr	Cytosol *un*stained	Very Poor
(b) Stain with anti Z IgG	Membrane bound vesicles stained	

Question 2. Skeletal muscle was stained by a standard periodic acid–Schiff procedure. Glycogen present in muscle fibres was strongly stained, but no mast cells were visualized in the connective tissue capsules. If, however, the oxidation time was

markedly extended, then mast cell granules were clearly seen. To what might this be due? Hint: consider the electric charges of glycogen, heparin, and periodate.

Question 3. A sequence polychrome stain devised by Shoobridge (1983) for routine pathological use uses four acid dyes in the following sequence: Naphthol Yellow S, Orange G, Acid Fuchsin, and Methyl Blue. Shoobridge has commented that 'when the dyes are used in reverse order, or other permutations are tried, staining is no longer so selective, and there are "muddy" colours present in many of the tissues'. Suggest why this is so. Hint: consider the probable cause of selectivity, and the relative sizes of the dyes.

5

Evaluating the significance of staining — and of non-staining

5.1 SCEPTICAL QUESTIONS AND THE EVERYDAY NEED FOR CONTROLS

As discussed in section 2.2, histochemical staining is carried out to answer questions such as: What is it? Where is it? How much of it is there? Staining outcomes must consequently be evaluated in the light of these objectives. To assist evaluation you can use a set of sceptical questions, which of course depend on whether the expected staining occurred or did not occur.

So now imagine you have carried out a staining procedure intended to demonstrate the presence of the intracellular enzyme Xase. If colouration of cells occurs, then the simplest explanation is that Xase was present at the site of staining. It is, however, useful to ask:

(1) Is the staining in fact due to the presence of Xase?
(2) Does the distribution of the stain correspond to that of the Xase?
(3) How is the intensity of staining related to the amount of Xase?

Such questions are sometimes addressed under the headings specificity, localization and quantification of staining respectively.

Now imagine no colouration arises in the cells when the staining procedure for Xase is carried out. An obvious interpretation is that no Xase was present. However, there are again useful questions to be asked, as follows:

(4) Is the lack of colouration due to non-staining of Xase?
(5) Has Xase, present in live cells, been lost or altered prior to staining?
(6) Has stain been lost from cells, or altered, prior to observation?

These questions lack such crisp labels as the first three. However, question (4) is sometimes addressed under the heading 'lack of sensitivity', and question (6) under the heading 'fading' or 'diffusion' of stain.

The range of control procedures which can usefully be used to answer such questions is wide; hence an overview is provided in Fig. 5.1. Each of the major topics

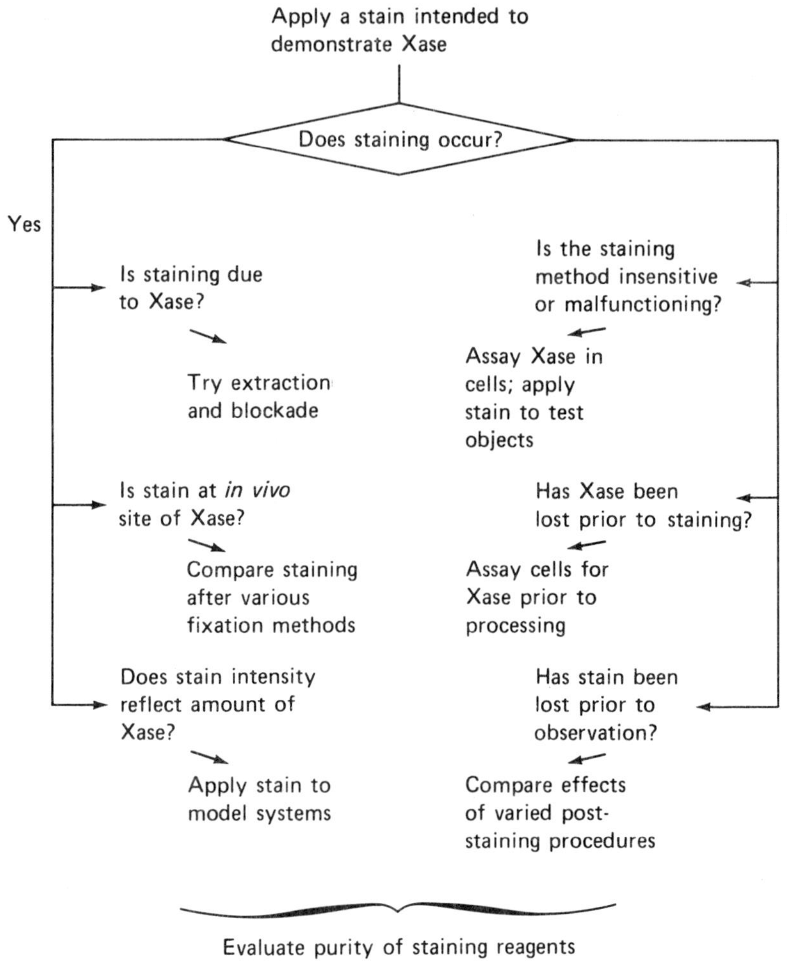

Fig. 5.1 — An algorithm summarizing control methods used in histochemistry. Note that the entity Xase could in reality by a chemical grouping, a particular compound, a biological structure, and so on.

mentioned in Fig. 5.1 are discussed below.

Note that benchworkers sometimes use control methods only retrospectively, that is after the expected results have not been seen. Such behaviour is over-optimistic.

5.2 EVALUATING STAINING

5.2.1 Assessing specificity using extraction, blockade, and more

How can we check that the supposed staining process is in fact the one occurring?

The two most elaborated methodologies (namely extraction and chemical modification of the substrate) are discussed first, followed by briefer mentions of various other tactics.

The use of *selective extraction of substrate* is conceptually simple. If a colouration is thought to involve Xase, then removal of Xase prior to staining should result in no colour occurring. If this expectation is not met, then evidently the staining is non-specific. Hence selective extraction is a useful tool, as long as procedures can be found which remove most or all of the Xase without altering anything else. There are two rather different approaches to this. The first is to use a solvent with a high affinity for Xase and a low affinity for everything else; this is often termed solvent extraction. The second is to selectively split the Xase into small molecular fragments, which are rapidly lost into an inert solvent. This selective fission may be achieved enzymically or chemically. Examples of these strategies are given in in Table 5.1. The extraction/fission distinction is not always clearcut. For instance, is the use of dilute acid to extract histones from chromosomes based on solvent selectivity, hydrolytic selectivity, or both?

Table 5.1 — Examples of selective extraction used as a control procedure

Type of extraction	Reagents used	Substrate removed
Solvent extraction	Hot chloroform–methanol	Lipids
Chemical fission	Cold aqueous perchloric acid	RNA
Enzymic fission	Aqueous amylase	Glycogen

A general problem with such extraction methods is non-selectivity; a few examples follow. Cold acetone is used to selectively extract hydrophobic lipids from cryostat sections. However, presence of traces of water, or extension of extraction times, or rises in temperature all facilitate simultaneous extraction of polar lipids. Even with enzymic extraction problems of specificity arise, such as the presence of traces of other enzymes. Moreover extraction procedures — whether using solvents or chemical and enzymic fission — can increase the staining of other substrates, probably owing to the non-specific removal of cellular components increasing staining rates. Note that as fixation, by definition (see section 2.3.2), gives rise to chemical modification of cellular bipolymers, extraction procedures are inevitably fixation-dependent. In fact this will not be a merely chemical effect, since fixation causes gross changes in permeability (see Fig. 2.2). Hence the entry of large molecules — such as extractive enzymes or stains — will also be influenced by the fixative.

The basis of the use of *histochemical blockade*, or blocking, is also straightforward. If colouration is thought to involve Xase, modification of the reactive portion of Xase prior to staining should eliminate or inhibit colouration. If this expectation is not met, then evidently the staining is non-specific. Blockade is a useful control process as long as the modification procedures do not themselves significantly alter the non-Xase tissue components. Blockades require reagents for the irreversible, quantitative and selective modification of Xase. This strategy can in fact be applied to a wide variety of histochemical techniques; some examples are given in Table 5.2.

Table 5.2 — Examples of histochemical blockade

Entity blockaded		Mode of blockade
Antigens		Labelled antibody staining can be blockaded by treating the section with an unlabelled specific antibody — which forms an antibody–antigen complex
Enzymes	(1)	All enzymes are inhibited by treating sections with protein denaturants such as heat or glutaraldehyde
	(2)	Particular enzymes are inactivated with specific inhibitors, e.g. myosin ATPase, unlike other phosphatases, is inhibited by Mg^{2+} and ADP
Amino acid residues	(1)	Amino groups can be eliminated by conversion to hydroxyl moieties using nitrous acid
	(2)	Carboxyl groups may be converted to non-ionic methyl esters using hydrochloric acid–methanol

Problems arising with the use of blockades largely parallel those previously discussed for extraction controls. Thus the chemistry of blockade reactions may have unsuspected non-specificities. An example is provided by the methylation of tissue carboxyls using hydrochloric acid–methanol as the esterifying agent (cited in Table 5.2). This reagent, if not freshly made up, contains small amounts of methyl chloride. Consequently methylation of tissue amines is a side reaction. Note that even if a blockade reaction is specific, its efficacy may vary from one tissue site to another, probably reflecting differential ease and degree of entry into different tissue sites of the blockade reagents (i.e. rate and affinity effects). A blockade of Xase can also increase the staining of other substrates, probably owing to increased staining rates following generalized losses of tissue components. Supporting this interpretation is the observation that such effects are more common when blockades involve extremes of pH, and when staining reagents are of large size. Fixative effects are again marked. One example involves the nitrous acid deamination of tissue proteins, as cited in Table 5.2. Such deamination is rapid following fixation in alcohol, but slow after formalin, probably owing to effects on the rate of reagent entry (compare Figs 2.2(a) and 2.2(c)). Finally consider a problem which cannot arise with extraction — namely that the blockade reaction may prove reversible under some circumstances. So blockade of an antigenic site with unlabelled specific antibody may be reversed in the staining bath, by replacement by labelled-antibody.

Another control procedure of general applicability is the omission of a reagent, or a reaction stage, from a multireagent or multistage procedure. Consider methods using Schiff's reagent after a prior aldehyde-generating step, for example the periodic acid–Schiff or Feulgen nucleal procedures. Omission of the pretreatment serves to check an occurrence of endogenous aldehyde, which could give false positive results. Another example is the omission of substrate in enzyme histochemistry, or in immunocytochemistry using enzyme-labelling. Occurrence of colouration under control conditions will indicate non-enzymic staining.

5.2.2 Assessing staining intensity — is quantification possible?

Consider the following problem. We stain sections from different portions of an organ, to demonstrate Xase. The different zones show marked differences in staining intensity. The question arises: do these differences reflect biological/pathological effects? And how big a staining difference is required to convince us that we are indeed seeing a biological change? So a more realistic heading for this section might have been 'does staining intensity reflect the amount of substrate present?'. However, quantification in the rigorous sense is sometimes what we are concerned with, especially when staining is used to answer 'how many?' rather than 'how much?' questions.

Examples of using staining intensity in an either/or manner — that is, intensely stained *versus* unstained, to assess the numbers of various entities — are familiar from haematology and cytology. Thus the relative numbers of basophilic, eosinophilic, or neutrophilic granulocytes in a blood sample can be determined in a Romanowsky stained smear by counting the numbers of cells containing violet, red or purple granules respectively. Similarly, in Papanicolaou stained vaginal smears, counts of superficial and basal cells are aided by the respective red and turquoise staining of these cell types. However, methods for estimating the amounts of specific compounds have also been developed for substrates as varied as acid phosphatase and DNA. Note also that between such questions as 'How many cells?' and 'How much enzyme?' come problems such as 'How many hormone receptors on the surface of a hepatocyte?', or, 'How many synapses in a ganglion?'.

A general problem arising with 'how much/how many?' questions is the need to standardize sample size. In microscopic practice this usually means standardizing sample thickness. When whole cells are investigated, as in cytology and haematology, then it is usually visually obvious if there is or is not a cellular monolayer. The same applies to investigations of whole organelles, such as chromosomes in cytogenetics. However, determining sample thickness is not so straightforward with 'slices' of tissue. The traditional phrase '6 μm sections were cut' usually describes hopes and the setting marked on a microtome, rather than a measured quantity. Thus intense and weak staining could merely reflect thick and thin sections. Section thickness is influenced, directly and indirectly, by differences in the type of fixation, processing schedules, and embedding media. Hence this property, although conceptually trivial, is often of considerable practical significance.

The influence of specimen thickness can also be more varied than the simple one discussed above. If you are investigating entities which are numerous, evenly distributed, and small compared to the thickness of the sample (for example macrophage lysosomes, in a 20 μm cryostat section) then comparisons might be made between different cells without gross errors. However, if the target entity is relatively large (for instance a motor neurone cell body in a 1 μm plastic section), or sparsely distributed (as are the nucleoli of such a neurone) or less symmetrically distributed (as is the glycogen within an alcohol fixed hepatocyte) then the difference in sample size resulting from section thickness variation can be confusing. Taking the examples above, then thick sections will give rise to more large diameter neuronal profiles, more neuronal profiles containing nucleoli, and more glycogen-containing hepatocytes than will the images from thin sections. Moreover, since thick sections reach staining equilibrium more slowly than do thin ones, errors may also arise due to non-staining of small entities located in the centre of thick sections.

There are of course further problems. How, for instance, will staining intensity be determined? Subjective assessment can sometimes be useful, as shown by the various 'scoring systems' used in clinical work. When objective measurements are used these may be made directly by microdensitometry of stained sections or smears, or by equivalent photometric measures of cells in flow cytometry; or the stain may be extracted from the sample and assayed. In any event, except for either/or judgements, successful assessment of staining requires that intensity of staining is proportional to the amount of substrate present. A few staining methods, such as the Feulgen nucleal procedure for DNA, are in fact stoichiometric but this is not essential. Thus quantitiative histochemical determination of proteins is possible, using acid dyeing under standardized conditions, although dye binding is certainly not a stoichiometric reaction. Such non-stoichiometric methods only yield relative values. To obtain absolute concentrations requires some kind of external standard. One approach is to compare histochemical staining with chemical or biochemical assays of the same tissue. Another is to stain model systems containing known amounts of substrate in parallel with the biological material. A widely used model system traps substrate (for example DNA or acid phosphatase) within a polymeric matrix, which is cast or sectioned to form thin layers. Another model system attaches the substrate (for instance antigenic peptide hormones) to the surface of small polysaccharide beads, of the sort sold as chromatographic support media.

5.2.3 Assessing localization of substrate and of stain

Here we ask the question: Is the stain observed down the microscope deposited where the substrate was to be found in the live cell? Alas not necessarily. Substrate may migrate from the *in vivo* site at any stage of processing up to and including staining. Moreover the stain itself may also become mobile, at any stage up to and including storage of mounted specimens. Examples of such artifacts are given in Table 5.3.

To avoid ambiguities of these types, various pre-emptive tactics are routinely adopted. *Post-mortem* changes are minimized by rapid dissection and/or application of fixative. Re-arrangement of substrate *by* fixatives is minimized by careful selection of reagents, and sometimes by using perfusion rather than immersion fixation, to reduce fixative flow across tissues. Deposition of stains in non-substrate sites is minimized by careful control of staining concentrations, pH, times, solvents and so on. Finally, redistribution of stain is minimized by use of dehydrating and clearing solvents, and mounting and embedding media, with low affinities for the stain.

It must be acknowledged, however, that actually *assessing* localization is both a less frequent and a more difficult activity. The only procedures at all commonly adopted are to compare the final images produced by different technical procedures: different fixative and processing regimes, different staining methods for Xase, and so on.

Table 5.3 — Examples of migration of substrates from their *in vivo* sites, and of stains from their sites of deposition

Technical stage		Example of artifact
Prior to fixation		Distribution of lysosomal contents can vary as a result of the way in which the animal was killed
During fixation		Cell contents can be precipitated along the side of the cell furthest from the incoming fixative, e.g. the 'alcohol flight' of hepatocyte glycogen
During staining	(1)	Lipid may be solubilized and relocated by the ionic surfactants present in Luxol stains
	(2)	Deposition of lead sulphide final reaction product of Gomori-type enzyme histochemical methods may occur in cell nuclei, not at the enzymic site
After mounting		Sudan dyes dissolved in fat droplets may form crystalline aggregates in tissues surrounding lipid

5.3 EVALUATING NON-STAINING

5.3.1 Assessing losses or alterations of substrate

When you stain for Xase and no colouration occurs, this may of course indicate a lack of Xase. However, one possible artifact is that Xase was present *in vivo* but has been lost prior to staining. As with the more minor forms of substrate mobilization, discussed in section 5.2.3, movements of substrate out of the tissues can occur at any point in the preparation of specimens for microscopy, up to and including staining. Loss can take an additional form, a loss of nature, owing to an alteration in chemical character. Thus Xase may be lost *in toto* from the tissue; or the atoms making up Xase may be retained but in some new configuration, usually with no enzymic character. Examples of losses of both kinds which occur during routine specimen preparation are given in Table 5.4. Unwanted (non-specific) losses can also arise

Table 5.4 — Losses of substrates — both of content and of character

Technical stage		Examples of losses of content (A), and of character (B)
Prior to fixation	(A)	Cells readily lose small, water-soluble molecules such as amino acids, glucose and potassium ions
	(B)	Many cytosol enzymes lose activity soon after removal of cells from the animal
During fixation	(A)	Proteins are extracted by aqueous osmium tetroxide, and lipids by aqueous glutaraldehyde
	(B)	Most fixatives, but especially reactive aldehydes, reduce or eliminate enzymic activity
During infiltration	(A)	Phospholipids are extracted into media such as wax, epoxy resins, and glycol methacrylate monomer
	(B)	Hot wax, and hydrophobic plastics, reduce or eliminate enzymic activity
During staining	(A)	Water soluble materials, e.g. glycogen and RNA, are extracted into aqueous staining solutions
	(B)	Enzymes are inhibited by many trapping and visualizing agents

during other preparative procedures, for example decalcification, and indeed, as described in section 5.2.1, during control procedures such as blockade and extraction.

5.3.2 Assessing staining failures with test objects

When you stain for Xase and no colouration occurs, yet another possible artifact is that Xase is present but that the staining procedure failed in some way. This can be checked with the aid of test objects. Many workers retain tissue blocks which give rise to 'good' staining of various substrates. A section from such a block is then stained together with each set of unknowns. If the test-slide fails to show colour, an immediate indication of technical failure is provided. Occasionally chemical or biochemical assay of processed tissue samples has been used to check if *any* Xase is present at the staining step. When you do conclude that the staining procedure for Xase has failed in some way, the next question must be — in what way has it failed?

One possibility is that the tissue sample, although containing Xase, is inhibiting the staining process. This can involve the action of an inhibitory substance. An example is provided by the staining of nuclear histones by certain acid dyes. This acidophilia is facilitated by a prior extraction of nucleic acids. Such pre-treatment is necessary as the negative charges of the nucleic acids otherwise mask the positive charges of the histones from interactions with the anionic dye. More commonly, inhibition is rate-mediated. Thus with a given staining procedure, the Xase in accessible tissue sites might stain strongly; whereas an equal amount of Xase at an equal concentration present within inaccessible sites might fail to stain. Examples of this include the numerous staining methods which require hydration-induced swelling of cells and tissues before the reagents can enter, such as the Alcian Blue staining of corneal connective tissue, and the Gram staining of bacteria. The effects discussed in this paragraph are strongly influenced by the mode of tissue processing adopted. Hence negative artifacts can arise if the methods of fixation, embedding and so on are not those assumed by the 'standard staining recipe' in use.

Another tissue-dependent factor involves our mode of visual perception. Consider staining tissue in which the same amount of Xase is distributed uniformly across the cells or is concentrated into a few organelles. Which will be most easily noticed?

Of course the source of the negative artifact may be the stain or the staining system. The 'error' may merely be that a staining procedure is too insensitive for the problem in hand. This can be assessed if a series of models, containing varying concentrations of the substrate — of the type described in section 5.2.2 — are stained. A chronic source of error is the purity, or rather the impurity, of staining reagents — this is discussed in section 5.4. Alternatively the staining procedure as used may not have been precisely as prescribed in the standing schedule. So check times, concentrations, pHs, solvents with care — most errors are trivial ones.

5.3.3 Assessing losses or alterations of stain

Stain may be extracted from the specimen, or may remain but be transformed into an invisible or ineffective form. Losses of one or other type can occur at any point up to and including microscopic observation. Examples of such negative artifacts are given in Table 5.5.

Although avoiding such artifacts in standard staining procedures is part of general laboratory lore, it is carried out in a conceptually fragmented way. As there is no systematic approach to assessing losses of stain, it is not only the beginner who can be puzzled. Gross fading or leaching is visually obvious, partial losses are more insidious. Perhaps the only generally applicable precaution taken in this area is the repetition of the staining after an interval, this providing a comparison of fresh and aged preparations.

5.4 EVALUATION OF STAINING REAGENTS

The uncertain quality of some staining reagents can be responsible for a variety of positive and negative artifacts. A reagent may not be what the label on the bottle says; or it may be only partly as labelled; and the diluent may actively influence staining. Moreover many reagents show considerable batch variation, even from a single supplier. Note by the way that samples of most types of staining reagents have been found wanting — from dyes to enzymes substrates, to visualization agents, to labelled antibodies, to chromogenic reagents. Evaluations of staining reagents come in two parts: knowing which substance is actually required; and knowing whether or not your sample of stain actually is the one required.

There are several reasons why the identity of a reagent may be uncertain. First, many commonly used staining reagents — dyestuffs, diazonium salts, coupling agents and so on — are, or derive from, industrial products. A compound used as a biological stain may be sold industrially under several trade names. Suppose a promising azodye enzyme histochemical method uses Nitrosamine Red as a visualizing agent. You will be out of luck if you search for this name in the catalogues of suppliers of stains and fine chemicals, although the substance is commercially available under the trade name Fast Red GG. A related issue is the use of non-systematic names, and cryptic initials. The need for such terminology is quite understandable. Would you prefer 2′,4′,5′,7′-tetrabromo-3′6′-dihydroxyspiro [isobenzofuran-1(3H)9′-[9H] xanthen]-3-one disodium salt or Eosin Y? Nevertheless trival names can confuse. Yes, I know what FITC and NBT are — but TRITC, and TCNBT? Moreover there are stains whose chemical identities have never been rigorously specified: from the traditional iron heamatoxylin to a new neurological tracer such as Nuclear Yellow.

To establish or specify the identity of many dyestuffs and related compounds the *Colour Index* is a valuable tool. This allocates a unique number and name to each dye structure of industrial significance. Thus the basic dye Methyl Green has also been sold as Double Green SF and Light green; however, confusion with similarly named acid dyes is avoided if the Colour Index number (42585) and generic name (Basic blue 20) of Methyl Green are specified. For dyes not so coded try using the Biological Stain Commission's monograph *Conn's biological stains* (Lillie, 1977). Nitrosamine Red, Fast Red GG, FITC and TCNBT do all appear in this handbook's index — although admittedly TRITC does not.

Let us now assume we know the desired chemical nature of a reagent. How might a sample we have purchased differ? In many ways — some of which are illustrated in Table 5.6.

As it is impracticable to assay or purify staining reagents coming into our

REED LIBRARY
STATE UNIVERSITY COLLEGE
FREDONIA, NEW YORK 14063

Table 5.5 — Losses of stains — both of content and of character

Technical stage		Examples of losses of content (A) and of character (B)
During post-staining rinses	(A)	Selective extraction of stain from fine fibres, e.g. from Aldehyde Fuchsin stained elastic fibres, or Acid Fuchsin stained collagen
During dehydration	(A)	Ready leaching of less hydrophilic dyes — such as piric acid and Pyronin — into alcohols
	(B)	Loss of metachromasia (with transformation to orthochromatic colouration) after dehydration
Following mounting	(A)	Leaching of non-ionic dyes (e.g. sudans) into resin media, and of acid and basic dyes into glycerine jelly
	(B)	Fading of many dyes and some inorganic pigments (e.g. PbS), both photolytically and chemically
During microscopic observation	(B)	Rapid fading of fluorescence during UV illumination of specimens stained with fluorochromes

Table 5.6 — Examples of problems arising from the use of 'impure' staining reagents

Type of 'impurity'	Example — and consequences
Stain specified is non-existent	Published methods for preparing Acetyl Sudan Black B are ineffective — samples so labelled contain only non-reacted Sudan Black
Nature of reagent required is unknown to anyone	SITS has been recommended as a fluorescent axonal tracer — but the active compound is actually a trace impurity of uncertain structure, occurring in only a few batches of SITS
Nature of reagent required is unknown to the user	Erie Black G, recommended as a granulocyte stain, seemed commercially unavailable — but use of *Conn's biological stains* indicated that the name is merely a synonym for the well-known Chlorazol Black E
Concentration of reagent very variable	Dye contents of commercial Pyronin Y samples varied from 13% to 65%: samples were ineffective in the Methyl Green–Pyronin (MGP) method if dye content was <50%
Colourless impurity actively influences staining	One Pyronin Y sample with >50% dye was nevertheless ineffective in the Methyl Green–Pyronin stain, since the colourless material was a surfactant
Coloured impurity influences staining	A commercial sample labelled Pyronin Y contained instead Rhodamine B & G — the former giving rise to generalized cytoplasmic staining

That the reagents in this table are dyestuffs merely reflects the present author's everyday concerns — enzyme substrates, labelled antibodies, and so on can also display unreliability.

laboratories, how are such pitfalls to be avoided? The only tactic of general applicability, as far as assessment is concerned, is to retain samples from effective batches to use as test stains. There are also a couple of general precautions. Dyestuffs can be purchased which have been certified by the Biological Stain Commission (BSC). Although the tests used by this independent body do not guarantee purity, they do assure users that dyes purchased will be effective for specified purposes. Moreover the BSC will trouble-shoot ineffective batches of stain for the purchaser. Another general precaution is to avoid prolonged storage of potentially labile materials, such as antibodies, enzymes, and colloidal metals.

5.5 INFLUENCES OF PRACTICAL STAINING AND PREPARATIVE VARIABLES

Consider a histochemical investigation aimed at assessing the amount and location of the intracellular enzyme Xase in a specimen.

Methods of tissue sampling and stabilization can have an immediate influence on whether or not Xase is retained within the specimen; and if it is, whether it is at the *in vivo* site; and if it is, whether the intensity of staining gives an indication of the amount of Xase present. Post-fixation processing steps (for example washing, dehydrating, infiltrating, embedding, dewaxing, etching — not to mention macerating, decalcifying, blocking and so on — can also influence the nature of the final image.

Preparation of optically thin layers — by procedures such as smearing and microtomy — often influence localization. The shape of cells can be influenced, as can sensitivity and quantitation. Such effects can arise directly, or via influences on rate-controlled staining mechanisms.

The staining process itself can also influence losses and localization of Xase. Hence successful staining outcomes often require precise control of practical variables — times, temperatures, concentrations and more. This may not be easy as the nature of the key variables is not always appreciated. One chronically confusing factor is the purity of the staining reagents. This in turn may be influenced by the conditions of staining, and of storage of stains; even pure reagents can decompose.

Post-staining processing procedures (e.g. dehydration, clearing, mounting, storage) can give rise to loss or modification or delocalization of stain from the site of deposition.

5.6 FURTHER READING

Assessing the significance of staining is central to our purpose and much critical work has been done on assessing particular methods. However, general accounts are rare. For an extended account of the topics discussed here — with many more examples, but perhaps less clarity — see Chapter 11 of my monograph (Horobin, 1982). To obtain details on specific topics the reader can also look in the indexes of previously noted histochemistry texts, under such headings as 'blocking methods', 'enzyme inhibitors', 'extraction' and so on.

5.7 TEST YOUR UNDERSTANDING

Imagine you have cut cryostat sections from a skeletal muscle biopsy. You wish to identify the type of muscle fibres present. To this end you stain one section using a formazan enzyme histochemical method for mitochondrial dehydrogenase. A serial section is stained by the periodic acid–Schiff procedure (PAS) to identify those glycogen-rich fibres using the glycolytic pathway.

(a) If the formazan method produces colouration of certain fibres, how would you check that this stain was indeed a result of enzymic activity?

(b) If such a check indicated the occurrence of non-enzymic staining, what processes might be implicated?

(c) Conversely, if such a check showed staining to be due to dehydrogenase activity, how would you determine that formazan was deposited close to the enzymic sites?

6

Influences of pre- and post-staining procedures

The final outcome of staining, as seen by the microscopist, depends not only on the staining procedures itself but also on how the specimen is treated before and after staining. Hence this chapter reviews the consequences of such procedures as fixation, embedding and mounting on the final visual image.

6.1 FIXATION

What is not present cannot be stained, and fixation certainly selectively influences those losses from live cells and tissues which occur during processing for microscopy. Such losses may be of entire molecules, of properties of molecules such as antigenicity or enzymic activity, or may be of substituent chemical groupings on the molecules. Losses were discussed in section 2.3.2, and will not be reconsidered here. Summarized below, however, are those fixative-dependent chemical and physical modifications of retained material which can directly influence the final pattern of colouration.

For instance the fixation process commonly alters retained biopolymers such that stain–tissue affinities are themselves altered. All fixatives denature proteins, exposing hydrophobic groupings, and so sections from fixed tissue blocks have a higher non-specific affinity for many organic, and hence hydrophobic, staining reagents than do unfixed cryostat sections. Of course many fixative effects are more specific, for example the alterations produced in the overall electric charge of retained biopolymers. Compare the three fixative agents ethanol, dichromate ion, and formaldehyde. Ethanol, although inducing protein denaturation, does not chemically alter ionizable groupings on the proteins. However, fixation with dichromate often results in an overall increase in cationic character, since the Cr^{3+} cations generated during oxidation of the tissue by dichromate are eventually bound to tissue carboxylate groups. Formaldehyde on the other hand reacts with tissue amines to form less basic groups, thus decreasing the cationic nature of the tissue proteins. These outcomes, differing as they do in terms of electric charge, result in differential affinities of ionic staining reagents, such as acid and basic dyes, tetrazolium salts, labelled antibodies, and so on.

Fixation also alters various physical characteristics of cells and tissues, which in turn influence staining outcomes. Such alterations may be molecular, e.g. cross-linking of tissue biopolymers by fixatives such as glutaraldehyde, or may be on a larger scale, such as the shattering of whole cells or tissue units by ethanol or other coagulative fixatives. Such physical effects can influence affinity; for instance cross-linking can inhibit the salt effects underlying the critical electrolyte concentration methods. Physical alterations also influence staining rates: crosslinking often reduces, and shattering increases, the ease of penetration of staining reagents into cells and tissues.

6.2 OTHER PRE-STAINING MODIFICATIONS OF THE SPECIMEN

In addition to fixation, a variety of technical operations alter the chemical composition of specimens prior to staining. Some procedures are intended to alter the chemical character of the specimen, for example control methods such as blockades and extractions (see section 5.2.1). Other procedures *aim* to alter physical or mechanical properties, for example decalcification and maceration. Moreover it is routine to change the solvent composition of specimens, by dehydration and clearing, before embedding in hydrophobic media. All such procedures can have unexpected effects on the staining process, by removing material from the specimen. Such unintended extraction can facilitate penetration of staining reagents, and hence affect rate controlled selectivities. Such extractions are most marked after the use of solutions of extreme pH, such as acid decalcifying agents and alkaline macerating solutions, or the use of elevated temperatures, as occurring with hot chloroform–methanol mixtures, or the use of enzyme preparations containing proteases, for instance those used in extraction controls. Influence on rate-controlled staining selectivity is most marked when the staining reagents are large, for example when using a dye such as Alcian Blue, or a labelled antibody.

6.3 EMBEDDING

All embedding media provide mechanical support to specimens during microtomy. However, media and modes of embedding vary widely, giving rise to a range of embedding-related influences on specimen colouration.

The specimen may be well infiltrated, or may be merely encapsulated within the medium. If well infiltrated, the medium may or may not have an influence on the chemistry of the specimen. Such influences largely depend on the temperature of infiltration, on whether the medium is hydrophilic or hydrophobic, and on whether the medium carries chemical groupings reactive with the common biopolymers. Obviously possible outcomes are complex. Fortunately the characteristics and consequences of routine types of embedding media and infiltrating regimes can be succinctly summarized, as is done in Table 6.1. In the light of this tabulation, some brief comments on the effects of embedding on staining are pertinent. The fact that exposure of specimens to elevated temperatures and hydrophobic media denatures proteins, and so reduces antigenicity and enzymic activity, is not surprising. Equally

Table 6.1 — Practical consequences of some routine embedding and infiltration regimes

Embedding medium: Chemical type	Embedding medium: Properties	Quality and temperature of infiltration	Thin sections possible	Hard specimens possible	Morphology well preserved	Easy stain penetration	Antigenically and enzymically active	Chemically little altered	General convenience
Ice (i.e. sections from frozen specimens)	hydrophilic, not reactive, soft	good, low				★★	★★★	★★	★
Paraffin wax	hydrophobic, not reactive, soft	good, elevated	★		★	★★	★	★	★
Plastics									
(1) Crosslinked epoxy or acrylic resins	hydrophobic, reactive, hard	moderate, elevated	★★	★	★★				
(2) Glycol methacrylates	hydrophilic, reactive, hard	moderate, can be low	★★	★	★★	★	★★	★	

The last seven columns are grouped under "Consequences for microtomy and staining[a]".

[a] Following tradition, the more stars the better.

understandable are the more general changes induced in biopolymers by reactive groups such as cyclic ethers and activated double bonds present in the monomers used in plastic embedding. The freezing and thawing inherent in cryotomy will tend to increase, and the elevated temparatures occurring during paraffin embedding and the thermal curing of monomers will tend to decrease tissue permeability. By now the implications for both affinity and rate-controlled selectivity will be clear.

Some media are not removed before staining, for example the plastics mentioned in Table 6.1 and the traditional medium, celloidin. Discussion of the effects of this on staining are deferred to Chapter 7.

6.4 POST-STAINING PROCEDURES

Stained specimens are routinely exposed to a variety of solvents. Typically a rinse follows staining, during which stain may be inadvertently redistributed or extracted. Sometimes such rinses are better regarded as unacknowledged steps in the staining process. Examples are the 'bisulphite' or 'running tap water' rinses following use of Schiff reagent in the periodic acid–Schiff and related staining methods. As these operations transform colourless specimens into ones which are magenta in colour, the term rinse is clearly a misnomer. Other post-staining washes remove stain differentially from the more permeable portions of the specimen, and in so doing increase selectivity. Such a wash is actually a rate-controlled de-staining, and would be better termed differentiation.

It is standard procedure to mount the stained specimen, that is to immerse it within a liquid film, which is usually then hardened to form a solid layer, the mountant. This is typically of high refractive index, to reduce light-scattering at the specimen–air interface. The mountant also serves to protect the stained specimen from environmental hazards such as abrasion and fading. Mountants must be selected to achieve these ends while avoiding artifactual alterations of the staining pattern. Therefore a widely used mountant is polystyrene, applied as a concentrated solution in a solvent such as xylene. After evaporation of this solvent, the polystyrene forms a solid film. Since the affinity of such a mountant and solvent for hydrophilic stains will be low, this medium is suitable for stains such as sulphonated acid dyes. Conversely polystyrene is unsuitable for mounting specimens with low molecular weight hydrophobic species, such as the non-ionic dyes precipitated as final reaction products in azodye enzyme histochemical procedures; they would be completely extracted from the specimen. Specimens stained with such reagents are therefore mounted in hydrophilic media, such as polyvinylalcohol or gelatin, applied as concentrated aqueous solutions.

Extraction of hydrophobic stains also occurs into the organic solvents used for dehydration, such as the alcohols. A commonplace example is the ease with which the typical basic dyes are lost during over-zealous dehydration. This indeed accounts in part for the continued use of that complex, unstable stain Aluminium haematoxylin. This metal complex dye is, unlike the routine non-mordant dyes, very hydrophilic — and so is insoluble in alcohol.

An example of a more complex problem is the metachromatic staining of glycosaminoglycans with dyes such as Methylene Blue. If specimens are well dehydrated after staining, metachromasia is reduced or lost as colouration reverts to orthochromasia, with consequent loss of selectivity. So a hydrophilic mountant is preferred in this case, even though such a medium may slowly extract water-soluble dyes from the specimen. Analogous dilemmas arise with the chemical fading of inorganic stains such as lead sulphide, with photochemical fading of a wide range of organic pigments and dyes, with the quenching of fluorochromes, and with solvent-facilitated diffusion of stains from their sites of deposition.

Such effects can also arise prior to mounting if, as is commonplace, stained specimens are dehydrated and cleared. The staining reagents most readily lost into the organic solvents used in these procedures will be relatively small and hydrophobic. Newcomers to practical histochemistry often wonder where the picric acid has gone from a picro-trichrome; or the Pyronin from a MGP-stained preparation.

6.5 FURTHER READING

There are many research papers discussing specific topics relevant to this chapter, from anti-fading additives for mountants to post-staining chemical modification of stains to prevent stain diffusion. In spite of this there are few general accounts of these technically significant phenomena. Certain of the topics were discussed in previous chapters of the present book, albeit from different viewpoints. Thus information concerning fixation and certain control methods (e.g. extractions and blockades) are found in sections 2.7 and 5.6 respectively. Perhaps the best single source for questions concerning embedding, fading, mounting and so on is the Lillie and Fulmer compilation (fourth edition, 1976), although it is very anecdotal.

6.6 TEST YOUR UNDERSTANDING

Question 1. Imagine you are trying to identify the type of glycosaminoglycans present in a region of loose connective tissue, by staining with Alcian Blue at various pHs. As a control procedure you extract a section with hyaluronidase solution prior to staining. Compared to the standard sections, staining of the connective tissue matrix in the extraction control was diminished. However, strong *nuclear* staining was seen in the control, though this did not occur in the standard sections. Disconcerted, you checked the effects of using hyaluronidase purchased from several different suppliers. All enzyme samples reduced staining of connective tissue matrix, but only two out of four induced nuclear staining. Now answer the following questions:

(a) Is the nuclear staining artifactual? If so, why does it occur?
(b) To what can you attribute the enzymic batch variation?

Question 2. Imagine you have identified macrophages present in the submucosa of mouse gut by using lysosomal acid phosphatase as a marker enzyme. You used two standard enzyme histochemical methods: a 'Gomori'-style metal-ion procedure in which the final reaction product was lead sulphide, and an azodye method, in which the final reaction product was a low molecular weight non-ionic dye. Microscopic observation immediately following staining showed that both methods gave rise to a 'granular' staining pattern, probably representing staining of individual lysosomes. Stained specimens were stored in various ways for several months (see Table 6.2 for details). Subsequently they were re-inspected, and staining artifacts found in several preparations. These were: diffusion of azodye into lipid droplets in the adjacent fat cells; fading, both of lead sulphide and azodye; and growth of an obscuring crop of fungal hyphae. In Table 6.2 write 'diffusion', 'fading' and 'fungus' in those boxes where these artifacts were *most likely*. Now justify your decisions!

Table 6.2 — Possible post-staining artifacts occurring after demonstrating acid phosphatase

Final reaction product	Unmounted sections		Sections mounted in PVA[a]	
	Stored in a cupboard	Left standing on a table	Stored in a cupboard	Left standing on a table
Lead sulphide				
Non-ionic monoazo dye				

[a] PVA is polyvinylalcohol.

7

Influences of specimen geometry on staining

So accustomed are we to looking down the microscope at and through thin specimens that it is easy to regard them as two dimensional. In fact the three-dimensional morphology of such preparations varies markedly. Specimens can take the form of continuous layers or dispersed fragments; they can be thick or thin; the proportion of their surfaces accessible to stain can be small or large; their surface profiles can be uniform or irregular. All these variations in geometry can influence the pattern of staining. Before considering these issues, note that certain related topics are discussed elsewhere: the shattering of cells and tissues by coagulative fixatives in sections 2.3.2 and 4.3.1, and the swelling of some tissue compartments owing to solvation in sections 4.3.1 and 4.6.

7.1 GEOMETRICAL CONSEQUENCES OF PREPARATION METHODS

The morphology of the thin specimens exposed to staining reagents predominently reflects the mode of specimen preparation, not the morphology of the native material. The point is expanded below, for several geometric parameters.

The specimen as stained may be a continuous layer, or a collection of separate fragments. Materials which are dispersed *in vivo,* for example the cells in peripheral blood or other body fluids, are often formed into thin layers by smearing. This gives rise to dispersed specimens of isolated individual cells. However, on occasion such cells are aggregated into clumps, which aggregates are then sectioned, yielding continuous cell layers in the form of thin slabs. Conversely, whilst solid organs may be sectioned to give thin slabs, nevertheless isolated cells are often removed from their surfaces and smeared to yield dispersed specimens. Examples are cytological preparations such as cervical smears and tracheal brush preparations. The discrete components of dispersed specimens are usually cells, hence their thickness is 'one cell'. However, there are exceptions, for instance chromosome splashes. Sectioned preparations range more widely in thickness, from less than 50 nm in ultrathin plastic sections used for transmission electron microscopy to more than 50 μm in sections used by neuroanatomists.

The proportion of a specimen's surface in contact with a staining solution is influenced by various technical factors, such as specimen thickness, and the presence or absence of occluding materials. Thus dispersed preparations, especially smears, are thin and possess a large surface area : volume ratio. Such preparations contain cell clumps, in which there is a local diminution of accessible surface. Continuous preparations, such as sections, inevitably possess less accessible surface, which further diminishes as the thickness of the section is increased. Occlusion of specimens occurs in two ways: firstly owing to superficial non-specimen layers, as with the gelatin of the photographic emulsion in autoradiographic preparations, secondly owing to interpenetrating embedding media which are not removed prior to staining, as with many plastic sections. In occluded specimens the accessible surfaces will be less than in equivalent naked preparations.

The uniformity of a specimen's surface profile is largely controlled by the factors already discussed. Dispersed specimens are uneven, with cell layers deposited from cell suspensions being most varied. Continuous preparations can also vary in their profiles. Sections are relatively uniform, although chatter of the microtome knife, and overall wedging of each section, can reduce this. Uniformity of profile typically increases from cryostat to paraffin to plastic sections, though even with the latter both chatter and scoring can occur. Moreover cryostat sections often display intracellular shattering, further increasing permeability. Staining of cryostat sections with a routine reagent such as Aluminium haematoxylin may be twice as fast as staining of otherwise comparable paraffin sections. Continuous preparations made by merely draping a sheet of unmodified tissue (for example omentum) over a microscope slide will not, of course, be uniform in profile.

These geometric variations are summarized, for the more common types of cell and tissue preparations, in Table 7.1. In the discussions that follow, the effects of these morphological differences on the staining process will be considered. For convenience the discussion first focuses on naked specimens, both continuous and dispersed, and then moves on to occluded preparations, in particular plastic preparations.

7.2 GEOMETRY AND STAINING OF NAKED AND OCCLUDED SPECIMENS

In general, specimen geometry influences staining via rate effects. Below, we first consider the differential influences on staining rates of dispersed and continuous specimen geometries, and then the effects of thickness and surface profile. Finally some influences of occlusion are described. Although plastic sections are a form of occluded preparation, they are—because of their increasingly general usage—discussed separately in section 7.3.

Uptake of staining reagents into dispersed preparations is usually faster than into sections. One example of this is provided by the Romanowsky–Giemsa staining of bone marrow cells. Smears show purple chromatin in 10–20 minutes, whilst sections take several hours to reach this colour. Another consequence of the thin and dispersed nature of dabs, smears and so on is that selective staining based on the relative rates of entry of two dyes, of the kind discussed in section 4.3.2, can be achieved with smaller inter-dye size differences than is possible with sections. For instance, with smears of cervical epithelial cells stained by the Papanicolaou method,

Table 7.1 — Geometrical features of some common types of thin preparations

Method of preparation	Characteristic geometrical features of preparations			
	Dispersion	Thickness	Accessible surface	Surface profile
Smears, dabs and cells deposited from suspensions	Mostly single cells, some clumps	Mostly one cell, except for clumps	High, lower in clumps	Uneven, from flattened to plump
Sections, embedding medium removed prior to staining	Coherent sheets *or* isolated elements	From less than one cell to many cells	Moderate, falls as sections thicken	Fairly uniform
Sections, embedding medium present during staining	Coherent sheets *or* isolated elements	From less than one cell to many cells	Low, owing to occluding embedding medium	Uniform only if infiltration is complete
Thin layers—any of above—beneath a non-specimen layer	Dependent on the nature of the specimen	Specimen can vary as above, superficial layer variable	Low, owing to occluding superficial layer	Dependent on the nature of the specimen

selective colouration of superficial and intermediate cells is achieved with the dye pair Eosin Y–Light Green. The anionic weights of these molecules differ by about 100 daltons. However, when mixtures of acid dyes are applied to sections, as in the trichrome methods, selective colouration requires the differences in anionic sizes to be at least twice as large.

With both continuous and dispersed specimens artifacts can arise from rate effects due to changes in specimen thickness. Thus several section stains claim to distinguish two biologically distinct states of various structural entities: active/inactive mitochondria, oxygenated/deoxygenated erythrocytes, and so on. Most such methods are actually demonstrating 'superficial' *versus* 'within the section' entities. Changes in section thickness alter the relative numbers of superficial and internal entities, and thus alter the proportions of the 'two states'. A variant that does occur is seen when internal entities merely fail to stain. This is most marked with thick sections, such as those used in neuroanatomical studies, when failure of nuclei to stain can result in underestimates of cell populations. Less misleading artifacts arise with trichrome staining, in which the thicker sections are, for the same reasons, understained by the larger dyes, shifting the colour balance.

Analogous artifacts also arise with dispersed specimens, for example within the cell clumps found among otherwise monocellularly dispersed cytological smears. The clumps are less easily penetrated by stains than are isolated cells, and hence may stain with the smallest dye, regardless of the identity of the cells. When cervical smears are stained by the Papanicolaou method for instance, clumps of basal cells can stain with Orange G, which will only stain highly keratinized dispersed cells.

Surface profiles can also influence staining rates. Smeared lymphocytes are typically flattened and smooth, whereas fibroblasts are more irregular and bumpy. Cytoplasmic colouration of these two cell types by the rate controlled Romanowsky–Giemsa stain is typically uniform and patchy respectively.

Relative *intra*cellular staining rates seem similar in both dispersed and continuous preparations: for example chromatin stains faster than ribosome-rich cytoplasms in smears and in sections. So the summary of relative staining rates of various entities, given in section 4.3.1, is probably applicable to naked preparations of any geometry. For the complicating effects of interpenetrating occlusions see section 7.3, below.

Whatever the specimen geometry, staining rates are reduced by superficial occluding layers. Examples of such layers include the gelatin of the photographic emulsion in autoradiographic preparations, and sputtered carbon coatings. As the consequences of such occlusion are not generally appreciated, artifacts can arise. Thus if radiolabelling is randomly distributed through the cells of a section, whilst staining of superficial (that is accessible) cells is more intense than that of deeper cells, then again an erroneous interpretation of 'two biological types of cells' can arise. If two staining reagents are used, entry of the larger reagent or the one with the highest affinity for the occluding layer, can be differentially retarded. For instance carbon coatings are extremely hydrophobic, excluding large, hydrophilic staining reagents, and strongly adsorbing hydrophobic reagents.

7.3 PLASTIC SECTIONS—A SPECIAL CASE OF OCCLUSION

7.3.1 Plastic? Why use that?

Plastic sections derive from specimens infiltrated with liquid monomers, polymerized to embed the specimens within solid blocks of polymer. Resulting sections are often exposed to stain without removing the embedding medium. Thus the plastic is an interpenetrating occlusion, between the stain and part or all of the specimen.

The chemical characteristics of plastics used in this way are very varied. However, all must permit thin sectioning and give good morphological preservation of the specimen. For transmission electron microscopy even ultrathin sections of the plastic must be stable in the electron beam; and for light microscopy the plastic must permit staining with a wide variety of reagents. The plastics widely used at present include crosslinked, hydrophobic epoxy resins used for electron microscopy, and acrylics, both crosslinked hydrophobic materials used for electron microscopy and the so-called hydrophilic glycol methacrylate used for light microscopy. Although the glycol methacrylate monomer is water-miscible, the polymer is best regarded as weakly hydrophobic, though certainly less so than the other resins mentioned. The various monomers can react with various chemical substituents on biopolymers, and hence plastic embedding can chemically alter embedded specimens.

7.3.2 The differential infiltration phenomenon

Despite the varying chemical and physical characteristics of plastic embedding media they all possess similar, and differential, infiltration behaviour. Several lines of evidence agree that the following components of biological specimens are poorly, if at all, infiltrated by the commonly used plastic embedding media:

basal laminae	mitochondria
cellulose cell walls	mucus granules
chromosomes	myelin sheaths

collagen and elastic fibres	nucleoli
the glycocalyx	red blood cells
glycogen	starch and zymogen granules

Other cytoplasmic and nuclear structures are, however, well infiltrated.

Hence, with the resins presently used it is the physically dense and/or hydrophilic specimen components which are poorly infiltrated. This is not surprising as the infiltrating mixtures are viscous and somewhat hydrophobic.

7.3.3 The consequences for staining

Even with reagents which can access all parts of an embedded specimen, staining of plastic sections is considerably slower than that of dewaxed sections of similar material. This overall non-specific rate effect is presumably due to occlusion of the specimen by the interpenetrating embedding medium. In keeping with this, slowed staining is most marked with large reagents. In a study of tissue processed conventionally into a commercial glycol mathacrylate medium, penetration of staining reagents whose ionic weights exceeded 1000 daltons was impracticably slow. Consequently it is not surprising that whilst staining of glycol methacrylate embedded specimens with large reagents—for example labelled immunoglobulins—can be achieved; such methods call for very specific processing and staining procedures. Is it perhaps those procedures which give incomplete infiltration and polymerization, and a low degree of crosslinking, which permit entry of the larger reagents? In keeping with this, the use of large reagents to stain specimens embedded for electron microscopy—that is in well infiltrated and highly crosslinked media—has never become a routine procedure. Indeed sceptics have suggested that such staining is never achieved with sections cut from well-infiltrated and cured resin blocks.

Since even the 'hydrophilic resin' glycol methacrylate is in fact slightly hydrophobic, the relative hydrophobicities of plastics and staining reagents also influence staining. If a hydrophilic resin such as glycol methacrylate is used, then small hydrophilic reagents of suitable affinity can stain both infiltrated and plastic-free compartments of a specimen. Very hydrophobic reagents will stain the more hydrophobic embedding media, and also the infiltrated regions of the specimen, leaving the plastic-free compartments unstained.

As mentioned in section 7.3.2, the typical monomers are excluded by both dense and hydrophilic components of the specimen. During staining, the dense structures often impede diffusion of staining reagents, and so stain slowly. However, once stained they will de-stain, that is differentiate, slowly also. The hydrophilic structures on the other hand, at least when using aqueous staining solutions which hydrate and hence swell specimens, favour rapid diffusion of staining reagents, and so will often both stain and de-stain quickly. This difference between dense and hydrophilic plastic-free compartments of plastic sections gives rise to the following simple rate-control mechanisms. Exposed to a single reagent, hydrophilic structures will tend to stain before dense ones. Once both types of structure have stained, the initial consequence of applying a second reagent (sequence staining) is to replace the first stain in the hydrophilic structures. This mechanism will be most important when staining hydrophobic plastic sections with hydrophilic reagents, and when there is no marked influence of affinity on selectivity. A dramatic example of such rate control occurs when aqueous solutions of Basic Fuchsin and Methylene Blue are used to stain semithin hydrophobic plastic sections. When Basic Fuchsin is applied before Methylene Blue, dense myelin sheaths are red and hydrophilic collagen fibres are blue, whereas when Methylene Blue is applied before Basic Fuchsin then myelin sheaths are blue and collagen fibres red.

These influences of plastic can all be considered as controlling access of reagents to the specimen. However, plastic embedding can also influence stain-tissue affinity, in particular by modifying the electric charges carried by a specimen's biopolymers. This is especially marked with epoxy monomers used for embedding specimens for electron microscopy. These reactive cyclic ethers make biopolymers less cationic. Consequently small basic (that is cationic) dyes stain plastic-free tissue components intensely, whilst comparable anionic dyes fail to stain. Indeed staining by acid (anionic) dyes requires an additional non-coulombic contribution to affinity, such as the stronger van der Waals forces arising with molecules possessing large conjugated systems.

7.3.4 Guidelines for staining plastic sections

These guidelines look at staining from a series of viewpoints. The plastic used, the staining reagents involved, the nature of the tissue component to be stained, and the processing procedures are all considered. Finally a summary of typical artifacts is given.

The nature of the plastic influences staining in various ways. To use the spectrum of stains developed for paraffin and cryostat sections, the plastic embedding medium of choice will be hydrophilic. For instance with appropriate allowance for slowed staining rates, reagents as varied as enzyme substrates, labelled antibodies, and non-ionic dyes can all give selective staining of tissues appropriately embedded in glycol methacrylate. However, if you wish to stain tissues embedded in hydrophobic and crosslinked resins for electron microscopy, then a limitation is the exclusion of hydrophilic and/or large reagents from plastic-infiltrated tissue components.

Within these overall constraints the physicochemical features of staining reagents have complex consequences, as summarized in Table 7.2. A property of some significance not mentioned in this tabulation is the pH dependency of the hydrophilic/hydrophobic character of some reagents. For instance the pK_as of many basic dyes are around neutrality. Consequently aqueous solutions of such dyes will contain hydrophilic dye cations plus the hydrophobic free bases. It is possible that such tautomerism, rather than merely charge sign, underlies the basophilia of epoxy-embedded specimens.

Differential infiltration makes certain tissue components differentially accessible to various types of staining reagent. It is only dense and/or hydrophilic structures which stain with hydrophilic reagents such as labelled antibodies. Even after embedding in glycol methacrylate, such plastic-free structures will stain more readily with these reagents than do infiltrated structures.

Several routine technical factors influence staining by altering or interacting with specimen geometry. If pre-embedding dehydration is incomplete, infiltration by hydrophobic monomers will be inhibited, and so occlusion will be reduced. Occlusion by plastic is also reduced when certain plasticizers are present in the infiltration

Table 7.2 — Some effects of reagent structure on the staining of plastic sections

Nature of the reagent	'Hydrophilic' plastics, i.e. slightly hydrophobic		Hydrophobic plastics, i.e. very hydrophobic and crosslinked	
	Effect on staining	Typical reagents	Effect on staining	Typical reagents
Size	Retards staining, especially with larger reagents	Alcian Blue and Sirius Fast Red	Reagent entry can be prevented	Colloidal gold and labelled antibodies
Hydrophilic/hydrophobic character:				
Hydrophilic	Infiltrated and plastic-free tissue stains	Methylene Blue and Ponceau 2R	Intense staining limited to plastic-free tissue	Basic Fuchsin and Fast Green
Slightly hydrophobic	Infiltrated and plastic-free tissues, also the plastic itself, stain	Carbolan Green and Janus Green	Both infiltrated and plastic-free tissues stain	Picric Acid and Phenylene Diamine
Hydrophobic	Only hydrophobic tissue components, e.g. lipids, stain	Oil Red O	Infiltrated tissue, tissue lipids, and the plastic itself stain; plastic-free tissue unstained	Sudan Black B and Oil Red O
Electric charge:			(assuming epoxy monomers)	
Anionic	Both infiltrated and plastic-free tissues stain	No restrictions except those noted above	Acidophilia generally suppressed; but plastic-free tissue stained by dyes with big conjugated systems	Amido Black and Coomassie Blue
Cationic	Both infiltrated and plastic-free tissues stain	No restrictions except those noted above	Intense basophilia of plastic-free tissue	Basic Fuchsin and Methylene Blue

mixture or are dissolved into the solid plastic. Such reagents can increase the permeability of plastic, for example addition of aniline or phenol to epoxy resins increases staining by various dyes. Certain influences of section thickness on staining were noted in section 7.2 above. Irregularities in sections arising during microtomy—such as chatter, scoring and wedging—have similar effects. Consider chatter in a glycol methacrylate section resulting from the knife blade striking a blood clot. If the section is stained with a trichrome mixture, the erythrocytes of the clot can be stained in parallel stripes of different colours. If occlusion effects are marked, as occurs if using hydrophilic staining reagents on specimens embedded in hydrophobic plastics, staining free floating sections can enhance intensity of colouration. This probably reflects the reagent entering via two surfaces not merely one; and perhaps also that sealing of small pores in the plastic occurs during hot-mounting of sections onto slides.

In conclusion, an outline of some staining artifacts characteristic of plastic sections is provided in Table 7.3.

Table 7.3 — Characteristic staining artifacts arising with plastic sections

Artifact	Possible explanation
Unexpected non-specific increase in staining of infiltrated tissue	Presence of plasticizers (in infiltration mixtures or staining solutions) increase the permeability of the plastic
Unexpected increases in staining by hydrophilic reagents of tissues embedded in hydrophobic resin	Inadequate dehydration inhibits infiltration by hydrophobic monomers: occlusion reduced
Staining of tissue-free plastic embedding media	(1) Hydrophobicity of staining reagent matches that of the plastic (2) Acidophilia induced by trapping of fatty amine catalyst in resin
Understaining by big reagents, e.g. larger dye of trichrome mixes	High molecular weight reagents are most sensitive to overall lowering of staining rates by occluding plastic
A tissue component stains in one site, but not in another	Non-staining site is neither dense nor hydrophic, so is well infiltrated and thus occluded
Failure to demonstrate antigenicity or enzymic activity	Denaturation of proteins by use of high polymerization temperature and/or hydrophobic plastics

7.4 FURTHER READING

A recent monograph on cytological staining (Boon and Druijver, 1986) explicitly discusses some of the consequences of smearing rather than sectioning cells. Otherwise the consequences of such technical procedures are rarely explicitly reviewed. Effects of the geometry of plastic sections are also little discussed, though see a recent book by Litwin (1985), and my earlier reviews (Horobin, 1982, Chapter 10; Horobin, 1983).

7.5 TEST YOUR UNDERSTANDING

Question 1. Thin layers of lymphocytes were obtained from a lymph node in three ways: by making dabs, by embedding in paraffin and sectioning, and by embedding in glycol methacrylate and sectioning. Each of these lymphocyte preparations were stained in two ways: by an enzyme histochemical procedure to demonstrate lysosomal esterase, and by Giemsa staining. Answer the following questions:

(a) How would the mode of specimen preparation influence the appearance (that is shapes, sizes and variability) of the lymphocytes?
(b) Which mode of preparation gives rise to the fastest, and which the slowest, demonstration of the esterase?
(c) Which mode of preparation gives purple stained nuclei least, and most, rapidly when using Giemsa stain?

Question 2. Sections of rabbit parathyroid gland were prepared as follows: (1) by embedding in paraffin wax and cutting 6 μm sections; (2) by embedding in glycol methacrylate and cutting 1 μm sections; and (3) by embedding in a hydrophobic resin and cutting 0.5 μm sections. Dewaxed paraffin sections and unetched plastic sections were stained with labelled goat anti(rabbit parathyroid hormone) immunoglobulin. The staining outcomes were as follows:

Preparation (1) showed diffuse and rather weak staining of secretion granules, and very weak staining of the surrounding cytoplasm.
Preparation (2) showed crisp and intense staining of granules, with no background staining.
Preparation (3) showed crisp but weak staining of granules, with no background staining.

It is probable that all three preparations gave staining artifacts. Suggest what these might be.

Part III

Applications of staining theory

8

Modes of action of standard staining techniques

Central to this chapter are mechanistic vignettes describing specific staining methods, or particular staining strategies. These are collected into several sections paralleling the historical sequence in which such methods first became popular, from 'procedures using dyes' to 'affinity histochemical strategies'. Each vignette is written to a common format, as below:

Objectives of the method or strategy.
Staining reagent's physicochemical features. Quantitative information concerning reagents is tabulated in Appendix 1.
Critical technical steps, including specimen preparation if mechanistically important, also *standardized* or *variant* methods, and comments on advantages and disadvantages.
Key questions needing answering.
Mechanism outline, including *small print* issues when illuminating, and *alternative views* when justice demands.
Implications for quantitative or standardized use, and general applicability of the procedure.
References specific to the method.

This continuity of layout assists comparisons between methods and methodologies, and clarifies their standard physicochemical roots.

8.1 PROCEDURES USING DYES

A common thread links these methods: all involve selective uptake and retention of dyes, without formation of covalent bonds or change in oxidation states. This distinguishes 'dyeing' from other procedures in which dyes are generated or used, for instance azodye and tetrazolium salt histochemical procedures, and demonstration of calcium ions using the chelating dye Alizarin Red S.

Unfortunately dyeing methods possess an image of mystery and magic for most benchworkers. Even writers of standard texts often regard such procedures as Best's carmine or Gomeri's Aldehyde Fuchsin, or the Gram, Romanowsky and trichrome stains, as of unknown mechanism, and hence of limited value. The selectivity and affinity of dyeing methods are, however, no more obscure than, say, enzyme histochemistry or labelled antibody staining: the same physicochemical factors are significant in all three methodologies.

8.1.1 Oversight and other general stains

Several classic stains visualize the overall morphology of specimens, to this end often generating constrasting colouration of common biological structures such as nuclei/cytoplasms, intermediate/superficial cells, erythrocytes/leucocytes and so on.

Gram staining

Objectives

To stain bacteria, distinguishing Gram-positive and Gram-negative species by their blue and red staining respectively. The method is also used to distinguish yeast species, and occasionally for staining histological sections.

Staining reagents

The routine procedure uses two cationic dyes, one blue (Crystal Violet) the other red (Safranin); whose cationic weights are similar, as are the sizes of their conjugated systems. Crystal Violet is slightly hydrophobic, and the addition of hydrophobic anions precipitates it from aqueous solution, which precipitates dissolve in organic solvents of high dielectric constant.

Critical technical steps

If both bacterial classes are present:

(1) Heat fix a smear preparation.
(2) Stain with Crystal Violet, after which both Gram-positive and Gram-negative bacteria are blue.
(3) Treat with an iodine/iodide (that is, triiodide) solution.
(4) Differentiate with acetone or ethanol — Gram-negative bacteria lose colour faster than Gram-positive species.
(5) Stain briefly with Safranin — Gram-negative bacteria become red, Gram-positive blue-black.
(6) Air dry and mount.

Many *variants* involve alteration of one or more steps. Crystal Violet can be replaced by other dyes whose triiodides are insoluble. Other anions, such as picrate, can replace triiodide. Many organic solvents are suitable as differentiators. Safranin can be replaced by a variety of red dyes, for example Neutral Red.

Key questions
Why do some bacterial species stain blue, others red? What is the purpose of the iodine/iodide? Are the alternative reagents understandable in terms of any proposed mechanism?

Mechanism
Bacterial staining by Crystal Violet and Safranin is basic dyeing; dyes are cationic, bacteria contain anionic DNA and RNA (see section 3.3.1). The iodine/iodide solution contains hydrophobic triiodide anions, which precipitate the poorly soluble dye triiodide salt. This dissolves in the differentiator more slowly and controllably than Crystal Violet in the chloride or acetate forms. Dye is removed more rapidly from the thin-walled Gram-negative species than from thick walled Gram-positive organisms. Safranin stains Gram-negative bacteria (thin-walled, no competing Crystal Violet) much faster than Gram-positive species.

Small print. Altering entry of differentiator or loss of Crystal Violet alters Gram staining. Methanol and other solvents of high dielectric constant disrupt Crystal Violet triiodide, and are good solvents for Crystal Violet. Consequently methanolic differentiation is too fast to control.

Alternative views. For a survey of an amazing variety of earlier opinions see Thompson (1966).

Implications
Gram staining depends on biophysical not biochemical differences between bacterial species. However, as discussed by Bartholomew, the method is capable of standardization.

References
The man who proved the case for rate control summarized his investigations (Bartholomew, 1962).

Haematoxylin and Eosin
Objectives
To display the cellular and histological features of specimens by staining cell nuclei, basophilic cytoplasms (and sometimes elastic fibres and mucins) blue-purple, with most other structures, from cytoplasms to collagen fibres, stained in shades of red.

Staining reagents
Eosin Y (that is 'Eosin') is a red anionic dye, with a substantial ionic weight and conjugated system. 'Haematoxylin' is unusual for a basic dye, being large and hydrophilic. The active component is a 2:1 aluminium–hematein metal complex dye, unstable under strongly acid conditions, with acid–base indicator properties. All 'Haematoxylin' solutions are mixtures of the Aluminium Hematein dye, aluminium ions, and hematein; many also contain unoxidized haematoxylin and the over-oxidized product oxyhematein. The composition changes with time, 'unripened' and 'over-oxidized' Haematoxylins both contain little Aluminium Hematein, and so are poor stains.

Critical technical steps

(1) Stain with Haematoxylin — nuclei, basophilic cytoplasms and so on become red-purple.
(2) Differentiate in acid alcohol, if needed.
(3) Stain with aqueous Eosin — after which most other structures are stained red or pink.
(4) Treat with aqueous alkali — which turns the Haematoxylin blue.
(5) Dehydrate through alcohols, and mount.

Variants. Eosin can be replaced by several red acid dyes. As a basic dye, however, Haematoxylin is unusual: it is not lost into the Eosin solution or the dehydrating solvents, and may be used after most fixatives. Though relatively complicated to prepare, store and use, its usual alternatives are other metal complex dyes. Some Haematoxylin variants stain elastic fibres and glycosaminoglycans, some do not.

Key questions

Why do nuclei stain blue, and cytoplasms red? Why is Haematoxylin not extracted into the Eosin dyebath or the dehydrating agent, but is lost into the alcoholic differentiating agent? Why do some Haematoxylin and Eosin variants have different staining patterns, with certain structures being unstained by either dye? Why is the method largely independent of fixative?

Mechanism

This is basic dyeing plus acid dyeing; the cationic dye stains the anionic glycosaminoglycans and nucleic acids; the anionic dye stains cationic proteins (see section 3.3.1). Haematoxylin is not extracted by the aqueous dyebath or the alcoholic dehydrating solvents because the dye's high ionic weight and large conjugated system inhibit diffusion (see section 4.3.1); in addition its hydrophilicity reduces alcohol solubility (see section 6.4). The acidic differentiator breaks the metal complex into fragments of low affinity, which are lost into the solvent.

Small print. Haematoxylin preparations display various salt effects: glycosaminoglycans of cartilage and mucins stain best at low salt concentrations; elastic fibres at high; nuclei and RNA-rich sites stain over a wide range of salt concentrations. Haematoxylin and Eosin is insensitive to changes in fixative, reflecting the large conjugated systems of the dyes, and consequent strong (non-coulombic) van der Waals binding to tissue elements. Staining of calcium deposits by Haematoxylin represents chelation of calcium ions by free hematein, present in most staining solutions. The step (4) 'blueing' exploits Aluminium Hematein's acid–base indicator properties. Eosin fails to stain elastic fibres and epidermal keratin because they are dense, a rate effect (see section 4.3.1).

Alternative views. Traditionally Haematoxylin was regarded as a mordant dye, binding to tissues via the metal ion (Baker, 1962).

Implications

Haematoxylin stains are relatively difficult to prepare, store and use, so their standardization would be difficult. For most applications alternative dyes need to be big and hydrophilic with large conjugated systems. Stains meeting this specification are either metal complex dyes themselves or, as with Cuprolinic Blue, expensive. Haematoxylin itself is better understood than many benchworkers think, or indeed than some standard books say.

References
A discussion of the special properties of metal complex dyes, such as Aluminium Hematein, is given by Marshall and Horobin (1973a). For accounts of salt-induced increases and decreases in staining by metal complex and other dyes see Horobin and Goldstein (1974) and Marshall and Horobin (1973b).

Papanicolaou stain
Objectives
To identify cell types in cell smears, using such features as degree of cytoplasmic keratinization and basophilia, and the nuclear morphology. Thus in cervical smears nuclei stain blue, and the cytoplasms of hyperkeratinized, superficial, intermediate and basal cells are orange, red, green and blue respectively.

Staining reagents
All Papanicolaou variants involve two anionic dyes, Light Green SF and red Eosin Y; some variants also use the anionic dye Orange G. Anionic weights, and sizes of conjugated systems, decrease from Light Green to Orange G. Staining solutions of the acid dyes always contain phosphotungstate anions, which with basic dyes form salts insoluble in either water or alcohol. Most variants use the hydrophilic metal complex Haematoxylin as a basic dye counterstain; for details see the vignette Haematoxylin and Eosin, above. A recent Papanicolaou variant uses Thionin (a small blue cationic dye) as a substitute basic dye. Papanicolaou stains traditionally contained the basic dye Bismark Brown, but this is not discussed since it has no influence on staining.

Critical technical steps
(1) Smear cells and, before they dry, fix with ethanol.
(2) Stain with Haematoxylin — which colours nuclei and basophilic cytoplasms blue.
(3) Stain with Orange G plus phosphotungstate, in alcoholic solution — after which cell cytoplasms are orange.
(4) Stain with a mixture of Eosin, Light Green and phosphotungstate, in alcoholic solution — which replaces Orange G with Eosin in superficial cells, and with Light Green in intermediate cells.

Some *variants* delete the Orange G step; some substitute Thionin for Haematoxylin in step (2). Descriptions of various standardized procedures are available (Boon and Kok, 1986).

Key questions
Why do cytoplasms of different cell types stain in such a range of colours? Why must smears be wet fixed? What makes it possible to substitute Thionin for Haematoxylin in step (2), even though Thionin is alcohol soluble and steps (3) and (4) use alcohol solutions?

Mechanism

In essence this is acid and basic dyeing; the anionic dyes stain cationic proteins, the cationic dye stains anionic nuclear DNA and ribosomal RNA (see section 3.3.1). The latter gives the cytoplasmic basophilia responsible for the staining of basal cervical epithelial cells. Other cytoplasmic colours are due to rate effects (see sections 4.2.1 and 4.3.2). In the cervix epithelial cells accumulate more keratin as they move away from the basal layer; in pathological circumstances cells can finally become hyperkeratinized. Since crosslinked keratin is poorly permeable, and since dye diffusion depends on both cellular permeability and dye size, the more keratinized cells stain with the smaller dyes. There are also differential non-coulombic affinity effects. Consider a highly permeable intermediate cell, rapidly penetrated by all three anionic dyes. If all remained staining would be muddy. However, the van der Waals attractions occurring with the dyes differ. It is the dye with the largest conjugated system, Light Green, which finally predominates in the cell, giving a crisp staining effect (see section 4.4.2). Hydrophobicity differences between the dyes are without consequence, since the solvents used are non-aqueous (see section 3.4.3).

Small print. Wet fixation is required as drying alters cellular morphology; and wet fixed nuclei best display the fine details of chromatin patterns. Moreover changes in morphology can alter staining rates (see section 7.2). Thionin, an alcohol soluble dye, can be substituted for Haematoxylin because of the phosphotungstate present in subsequent staining solutions; this polyanion converts the basic dye to an extremely insoluble salt. The phosphotungstate also plays a role in regulating Eosin–Light Green selectivity.

Implications

In spite of complications steps towards standardization of several Papanicolaou variants have been taken. The Thionin variant will probably prove useful, as this dye can be obtained pure. The importance of rate control suggests the need for standardization of specimen preparation methods; but the procedures will remain influenced by specimen geometry — for instance, cell type.

References

The importance of staining rate and the role of phosphotungstate is discussed by Boon and Druijver (1986). Use of Thionin and the role of specimen preparation on cellular morphology are considered in Boon and Kok (1986).

Romanowsky–Giemsa staining

Objectives

Identification of cell types in peripheral blood, bone marrow and cytological biopsies; typically in smears, sometimes in sections. This is achieved by colouring nuclear chromatin and some cytoplasmic granules purple, basophilic cytoplasms and nucleoli sky blue, erythrocytes and some granules red.

Staining reagents

The active components of Romanowsky stains are Eosin Y, a red anionic, rather hydrophobic dye with a substantial ionic weight and conjugated system; and Azur B, a small, blue, cationic dye. Although pure Azur B is now available, commercial Romanowsky stains — perhaps labelled Giemsa or Leishman's — still often contain numerous Methylene Blue homologues.

Critical technical steps
(1) Smear cells, air dry, and fix with methanol.
(2) Stain with a, largely aqueous, solution of the Azur B–Eosin mixture — giving the staining pattern noted above.

Using pure dyes a *standardized* Romanowsky procedure has been developed by the International Commission for Standardization in Haematology. This eliminates the need for most *variant* methods, which arose due to variations in Azur B content of polychrome Methylene Blues. Variant methods for section staining are still necessary. Such procedures use long staining times, followed by differentiation, typically in aqueous acetic acid, to obtain the specified polychromasia.

Key questions
What factors cause the polychromasia? In particular, why does nuclear chromatin stain purple but basophilic cytoplasms stain blue? Is methanol fixation of smears obligatory? Why does section staining need prolonged staining times, and acid differentiation? Is the Azur B–Eosin Y combination obligatory?

Mechanism
Initial colouration is acid and basic dyeing. Anionic Eosin stains cationic proteins pink; cationic Azur B stains anionic DNA and RNA pale blue (see section 3.3.1). Subsequently Eosin interacts with the Azur B bound to certain sites, yielding purple Azur B eosinate. The intracellular formation of this insoluble salt parallels its slow precipitation from the aqueous staining solution. The more permeable lysosomes and nuclear chromatin permit Eosin to enter quickly, and so to stain purple; the less permeable ribosomes exclude Eosin, and so remain blue (see section 4.3.1).

Small print. As precipitation of the purple salt is not an equilibrium process, it is not self-limiting. Hence visibility of small purple neutrophil granules depends on how much precipitation occurs, namely on such technical influences as staining time. Consequently any neutrophil can appear to be agranular, to be normal, or to contain toxic granules. Methanolic fixation of smears is necessary for purple staining; formaldehyde fixation yields blue chromatin. This is because alcohols, unlike formaldehyde, are coagulative fixatives which shatter cells, and so speed up staining rates (see section 6.1). The extended staining times required to develop the colour purple when using tissue sections reflect the relatively rapid staining rates of smears (see section 7.2). The acidic differentiator is more selective, as the low solubility of Eosin under acidic conditions permits Azur B to be removed preferentially. Use of Eosin is not obligatory, some other red hydrophobic acid dyes also yield insoluble purple salts; with larger dyes staining times may need extending (see section 4.6). For reasons not currently understood, Azur B is unique. If Methylene Blue, differing by only a single methyl group, is substituted the stain gives rise to blue, not purple, nuclei.

Alternative views. It has been argued that owing to affinity factors an Azur B–Eosin complex is formed preferentially in the presence of specific biopolymers, for example with DNA but not with RNA (Wittekind and Gehring, 1985).

Implications
As Romanowsky staining is not an equilibrium process quantitation is tricky; however, the method can certainly be standardized, and a rational trouble-shooting checklist established.

References
A standardized Romanowsky staining procedure has been specified by a working party of the International Committee for Standardization in Haematology (ICSH, 1984). The rate control model is discussed, and a trouble-shooting guide set out, by Horobin and Walters (1987).

Alkaline Toluidine Blue for plastic sections
Objectives
To display the cellular and histological features of semi-thin sections of epoxy-embedded specimens, typically as a preliminary to electron microscopic examination. Alkaline Toluidine Blue gives staining patterns analogous to the electron microscope image; strongly stained entities include heterochromatin, red blood cells and secretion granules.

Staining reagents
Toluidine Blue is a small, slightly hydrophilic, cationic dye. Variant methods use other cationic dyes (such as Crystal Violet, Methylene Blue, Safranin and Thionin) varying in size and hydrophobilic/hydrophobic character.

Critical technical step
Assuming epoxy-embedding:

(1) Stain by covering sections with aqueous, alkaline Toluidine Blue, and heating to 60–70°C on a hot plate or over a flame — after which the staining pattern is as noted above.

Higher temperatures are achieved in a *variant* method by adding polyethylene glycol to the staining solution and heating above 100°C.

Key questions
Why does alkaline Toluidine Blue give a staining pattern analogous to the image seen in the electron microscope? Why is a hot, alkaline solution needed?

Mechanism
The staining is essentially basic dyeing: the cationic dye binds to anionic nucleic acids, and also to proteins since they too are anionic under alkaline conditions (see sections 3.3.1 and 3.6). However, this alone would give rise to uniform staining,

which does not occur. The observed selectivity is due to the differential infiltration of the specimen by the hydrophobic embedding medium. Well-infiltrated components then exclude the hydrophilic dye, whereas poorly infiltrated or non-infiltrated components allow dyeing to occur (see sections 7.3.2 and 7.3.3). The similarities of the pictures produced by Alkaline Toluidine Blue to the images seen in the electron microscope are because dense structures are clearly pictured by both methodologies. In light microscopy dense structures exclude plastic, and allow dye penetration; in electron microscopy dense structures give rise to electron scattering.

Small print. Several factors may aid penetration of stain into the relatively impermeable plastic sections. Elevated staining temperature can increase rates of dye diffusion. Hot alkali can etch the sections. Moreover the alkaline conditions partially convert cationic dyes into their non-ionic free-bases, which hydrophobic species can dissolve in the hydrophobic embedding medium.

Implications

This is a convenience method, fast and simple, and applicable to any section, regardless of the specifics of tissue, resin formulation, or modes of infiltration and polymerization. The corollary of this flexibility is of course a non-standard character — since its advantages revolve around the ease of varying time in the stain.

References

Various effects discussed here have been reviewed in some detail elsewhere (Horobin, 1983).

8.1.2 Stains for connective tissue fibres

Methods are described here for two protein fibres of routine concern to histophatologists, namely collagen and elastic fibres. The Trichrome staining systems, considered here because of their ability to selectively stain collagen, are also used as oversight stains. Stains for other components of connective tissue are discussed elsewhere: for glycosaminoglycans, and hence for heparin containing mast cells, see section 8.1.4.; for neutral lipids, and hence for lipocytes, see section 8.1.5; and for the demonstration of acid phosphatase, a marker enzyme for lysosomes and hence macrophages, see section 8.4.2.

Dyeing collagen fibres — the Van Gieson and other Trichrome stains

There are many Trichrome stains, as outlined in the *Tailpiece* to this vignette. The Van Gieson procedure is taken as a type example because of its efficacy, wide usage, and relative simplicity.

Objectives

To demonstrate collagen fibres, and sometimes other components, by their red colouration, with nuclei stained black or blue, and other tissue elements shades of yellow.

Staining reagents

The original procedure used a mixture of the following acid dyes: first, the yellow dye Picric Acid with a low anionic weight, small conjugated system, and slightly hydrophobic nature; second, the red dye Acid Fuchsin with a higher anionic weight and larger conjugated system, which is extremely hydrophilic. A widely used variant substitutes the very large Sirius Red F3B for Acid Fuchsin. The original procedure used as a nuclear counterstain Iron Haematoxylin. This large, black, cationic metal complex dye is fairly stable under strongly acidic conditions; a property shared by a modern alternative, the iron complex of Celestine Blue.

Critical technical steps

(1) Stain with Iron haematoxylin — nuclei and basophilic cytoplasms become black.
(2) Differentiate in acid alcohol.
(3) Stain with an acidic, aqueous Picric acid–Acid Fuchsin mixture for a short period — producing red collagen fibres, with other structures coloured yellow.
(4) Dehydrate quickly through the alcohols.

Most fixatives can be used, but the red–yellow colour balance is fixative-dependent. Changing fixatives may require changes in staining time, or dye concentration. Many workers use a *variant* in which Sirius Red F3B replaces Acid Fuchsin in step (3), to improve resistance to fading. Some workers replace Iron Haematoxylin with the technically more convenient Iron Celestin Blue/Aluminium Haematoxylin sequence. For comments on other Trichrome variants see the *Tailpiece*.

Key questions

Why does collagen stain red, but cytoplasms yellow? Why does overlong alcoholic dehydration remove Picric Acid, but leave the Acid Fuchsin? Why are the metal complex dyes Iron Haematoxylin and Iron Celestin Blue used? Why is the colour balance influenced by fixation?

Mechanism

Fundamentally this is acid and basic dyeing. The anionic dyes Acid Fuchsin and Picric Acid stain cationic proteins, the cationic metal complex dye stains anionic DNA and RNA (see section 3.3.1). The fact that staining due to the anionic dyes is not a uniform orange is due to rate effects. In the short staining time used, the large Acid Fuchsin anions only have time to enter the most permeable structures, including the collagen fibres. Picric Acid does not accumulate in the collagen however, since the dye has a smaller conjugated system and so forms weaker van der Waals attractions than the red dye. The Picric Acid that does enter leaves during rinsing and alcoholic dehydration, owing to its small size (see section 4.3.2). The two anionic dyes have another marked difference: Picric Acid is slightly hydrophobic, Acid Fuchsin is extremely hydrophilic. Consequently over-extended alcoholic dehydration selectively removes Picric Acid from the section. The iron metal-complex dyes are used since they are more stable at low pH, and so resist extraction into the picro-Acid Fuchsin solutions.

Small print. On occasion, and after certain fixatives, structures additional to collagen stain red: for instance the zymogen granules of salivary glands. These contain glycoproteins which will swell in aqueous media. Such swelling also occurs in collagen, especially at the low pH generated by the Picric Acid. Permeability is also markedly influenced by fixation. A coagulative fixative increases overall staining rates, a crosslinking fixative reduces them; giving rise to redder or yellower colours respectively (see section 6.2). Owing to such rate sensitivity, Van Gieson's stain is strongly influenced by plastic embedding (see section 7.3.3). If the staining time of plastic sections is not extended, or rates increased by heat or use of plasticizers, then the colour balance is yellow.

Alternative views. One school of thought attributes the selectivity of Trichrome stains to specific affinities (for example, Lillie, 1977).

Implications

The rate control mechanism means that standardized staining solutions or procedures must be flexibly defined, since staining times and/or dye concentrations must on occasion be adjusted to cope with changes in fixative, embedding medium, section thickness and so on. Incidentally Acid Fuchsin is an innately heterogeneous material, always containing components of varying degrees of sulphonation and methylation: dye standardization of Sirius Red F3B or an analogue would be more feasible.

References

An elegant account of the rate control of Trichrome stains has been given by Baker (1958).

Tailpiece

The three major classes of Trichrome stains are all rate-controlled. There are many one-bath Trichromes which can be regarded as variants of the Van Gieson method, since they use mixtures of two or more acid dyes. A simple structure–staining correlation applies to such methods: the anionic weights of the dyes always differ by at least 200 daltons, and it is always the larger dye which stains collagen fibres (Horobin and Flemming, 1988). When one dye of the pair is not Picric Acid, then a mineral acid is always added to lower the pH. The sequence Trichromes form a second group, in which two or more anionic dyes are applied in successive dyebaths. Such staining is controlled only by tissue permeability; the first dye to be applied is always replaced most quickly in collagen fibres, regardless of relative dye anionic weights. Finally there are polyacid Trichromes, of which Massons's is a classic example, which make use of phosphomolybdic or phosphotungstic acids. These methods exploit a polyacids ability to act as a 'large colourless acid dye' (to quote Baker, 1958), and hence to selectivity differentiate or block collagen fibres. The practical use of Trichrome stains of all kinds has recently been clarified, including setting out a trouble-shooting system (Shoobridge, 1983).

Stains for elastic fibres

Several procedures are used, for instance Aldehyde Fuchsin, Orcein, Resorcin Fuchsin, Sheridan's Crystal Violet, and Verhoeff's Iron Haematoxylin. All can be rationalized by a single mechanism, so this vignette discusses these classic stains — along with Congo Red, Direct Blue 152 and Durazol Blue 8G, as infrequently used but illuminating additions — as if they were variants of a single procedure.

Objectives

To stain elastic fibres, and as little else as possible.

Staining reagents

The chemistry of elastic fibre stains is very varied. Some dyes are anionic (for example Congo Red), some cationic (such as Aldehyde Fuchsin); some hydrophilic (for instance Direct Blue 152), some hydrophobic (as is Sheridan's Crystal Violet); some carry hydrogen bonding groups and some do not (Verhoeff's Iron Haematoxylin and Durazol Blue 8G respectively). There is, however, one similarity underlying this diversity, all the stains have large conjugated systems.

Critical technical steps

An outline comes first, and details of the 'variants' follow.

(1) Optional oxidation with potassium permanganate — this enhances, but is not essential for, staining of elastic fibres.
(2) Stain with dye dissolved in a partially non-aqueous solvent, containing acid *or* alkali — after which the elastic fibres, and sometimes additional structures, are stained.
(3) In the latter case, differentiate.

Of the '*variants*' mentioned above, only Aldehyde Fuchsin and Resorcin Fuchsin commonly use an oxidation step. The usual non-aqueous solvent in step (2) is ethanol, present in proportions varying from 95% to 70% by volume; other solvents are sometimes used, such as dimethylsulphoxide. The adjustment of acidity/alkalinity follows a simple rule: acid dyes are applied from alkaline dyebaths, basic dyes from acid dyebaths. The existence of several methods for staining elastic fibres can be attributed to a variety of advantages and disadvantages. Thus Verhoeff's Iron Haematoxylin is fast acting, and gives a contrasty and permanent staining; but it must be freshly prepared before use, and needs differentiation under the microscope. An Orcein staining solution is easy to make up, and stable once prepared; but the purity of commercial dye samples varies markedly, leading to erratic staining patterns. And so it goes.

Key questions

Why do only elastic fibres stain, and why should so many different dyes stain them? What is the role of the prestaining oxidation? Why is fixation typically unimportant?

Mechanism

The crucial factors in these stains could almost be called an 'unstaining mechanism'. Dyebaths are substantially non-aqueous, reducing the contribution of hydrophobic bonding to affinity (see sections 3.4.3 and 4.6). The use of acid conditions with cationic dyes, and alkaline conditions with anionic dyes, ensures that coulombic effects reduce the affinity of normal proteins (see section 4.2.1). The presence of electrolyte, namely acid or alkali, inhibits binding of cationic dye to tissue polyanions, such as DNA and RNA (see section 3.6). The need for large conjugated systems is now apparent: to ensure that van der Waals attractions are strong enough to give rise to staining.

Selectivity arises from the unusual chemistry of elastin, the major protein in the fibres. Elastin has very few ionic side chains, so it does not — unlike normal proteins — repel the dyes coulombically. Moreover elastin's hydrophobic character gives rise to hyrophobic bonding, though probably only with the more hydrophobic dyes. Some dyes lacking hydrogen-bonding groups do stain elastic fibres, though it is possible that uptake of some other dyes is facilitated by formation of such bonds. As noted above, the electrolyte concentration in the staining solutions is high. This gives increases, as well as decreases, of staining. In particular, elastic fibres stain more intensely with cationic dyes under acid conditions when salt is added.

Small print. Enhancement of staining with cationic dyes by prior oxidation is due to formation of additional anionic groups. For instance the fibrillar component of elastic fibres contains a protein rich in thiol groups, which permanganate would convert to sulphonate. The lack of fixative effects is congruent with the role of van der Waals attractions; whatever the nature of fixed proteins such attractions will still occur.

Alternative views. Acidophilia, formation of covalencies, and mordanting have all been proposed as reponsible for the staining of elastic fibres, as reviewed in the paper cited below.

References

A discussion of the general mechanism is available, which account also summarizes structural information on classic elastic fibre stains (Horobin and Flemming, 1980).

8.1.3 Chromosome staining methods

Two rather different examples are discussed here. One procedure demonstrates chromosomes in many types of whole-cell preparations: cell suspensions, cell smears, and tissue fragments. The other method identifies individual chromsomes in metaphase spreads produced from suspensions of cultured cells. Chromosomes can also be stained using the Feulgen nucleal procedure, described in section 8.3.1.

Acetic Orcein stain

Objectives

To stain chromosomes, with minimum background, in very varied preparations, including whole cells and tissue fragments.

Staining reagent

Orcein contains a number of dyes, of different properties. The ones most probably active in this stain are large, slightly hydrophobic cationic dyes, with substantial conjugated systems. The substitute dyes Lacmoid and spirit soluble Nigrosin contain components with similar properties.

Critical technical steps

(1) Immerse specimen in an aqueous acetic acid solution of Orcein, for variable times depending on the specimen.
(2) If the specimen is thick, crush gently under a coverslip until a single cell layer is obtained — giving red staining of chromosomes of dividing cells, and of the chromatin of interphase nuclei.

There are a number of *variants*. For instance specimens may be fixed, before or after staining. Acidification may be achieved by the addition of hydrochloric acid or lactic acid to the staining solution. Staining may occur in more than a single step, one of which may involve heating. Certain dyes can be substituted for Orcein, for instance Lacmoid and spirit soluble Nigrosin.

Key questions
What causes the dramatic contrast between chromatin/chromosomes and all else? Why can the method be applied to specimens as varied as chromosome splashes, whole cells, or even tissue fragments?

Mechanism
This is essentially basic dyeing, of anionic DNA with the cationic dye species. Staining of protein does not occur, since at the low pH of the dyebath the protein will also be cationic (see section 3.3.1). The high concentration of acetic acid present assists selectively in additional ways. DNA is insoluble in acetic acid, whilst typical proteins swell grossly and are substantially extracted. Moreover RNA is more labile to acid hydrolysis than is DNA, and so cytoplasmic basophilia will be preferentially reduced. The swelling of proteins also assists penetration of dye, and the flattening of thick specimens such as tissue fragments. The nature of the dyes suggests that hydrophobic bonding and van der Waals attractions will contribute substantially to affinity (see section 3.6).

Small print. Most acids will cause these effects, which are moreover exacerbated at elevated temperatures. However, heating of staining solutions must be carried out with caution, since total hydrolysis of DNA, or indeed maceration of the cells, can result.

Implications
Although the gross swelling induced by acidic solvents permits application to a wide range of specimens, this make it necessary to tailor precise times and temperatures of staining to particular specimens to avoid under- or over-treatment.

References
The range of application of acetic Orcein is demonstrated by looking in the index of texts discussing chromosomes (such as the manual of MacGregor and Varley (1983)).

Quinacrine banding of chromosomes
Objective
To identify individual chromosomes by the patterns of transverse bands (Q-bands) formed on the chromosome arms.

Staining reagent
Quinacrine is a slightly hydrophobic, fluorescent dye, of moderate size, as are the alternative dyes Hoechst 33258 and Quinacrine Mustard. These dyes are largely cationic under slightly acid conditions.

Critical technical steps

(1) Fix chromosome preparations, for example in alcohol.
(2) Stain in a weakly acidic aqueous solution of Quinacrine — giving fiuorescent Q-banding.
(3) Mount in water, and observe/photograph using minimal exposure to UV.

The Q-banding pattern may also be obtained with Hoechst 33258 and Quinacrine Mustard. This method is technically simple; however, fluorescence fades rapidly under ultraviolet illumination. In contrast Giemsa-banding is technically trickier, and less consistent, but gives coloured and permanent preparations.

Key questions

Why is fluorescence confined to the Q-bands? Why mount the specimens in water? Why go to the trouble of using a fluorescent dye, especially when it fades so quickly?

Mechanism

Uptake of Quinacrine is basic dyeing; under the acidic conditions of application the dye is cationic, whilst DNA is anionic. However, the Quinacrine is taken up along the whole length of the chromosomes; the banding selectivity is due to the quenching of fluorescence occurring differentially along the chromosomes. At least in part this reflects the distribution of AT versus GC base pairs; since fluorescence is enhanced by the more hydrophobic character of the AT bases.

Small print. Quenching of fluorescence is also influenced by the mounting medium. A hydrophobic medium non-selectively enhances fluorescence, so banding is not seen as whole chromosomes stain. Other fluorochromes reported to give the same, or similar, banding patterns have conjugated systems of similar size, and have similar hydrophobicities. Fluorochromes are needed to effectively visualize the Q-banding pattern. Whilst similar banding patterns have been seen with certain non-fluorescent basic dyes (for instance Neutral Red and Paraosanilin), the bands resulting have proved too weak for routine observations. In this context it is of interest that the purple staining produced by G-banding procedures are generated by a non-equilibrium process (see section 8.1.1) — and their intensities are thus not limited by the low mass of chromosomal material comprising the banded regions.

Alternative view. The original workers, using Quinacrine Mustard, expected dye binding to be due to the formation of covalent bonds with specific bases; however, they noted that the non-reactive Quinacrine itself gave the same banding pattern (Caspersson *et al.*, 1970).

Implications

This methods displays chromosome banding using a technically simple procedure, at the price of requiring a fluorescence microscope, and of not providing permanent preparations.

References

A general technical and mechanistic sketch is given by MacGregor and Varley (1983). The effects of mountant, and the demonstration of differential quenching, are two of Sumner's many contributions to understanding chromosome banding (see Sumner (1977) and Sumner *et al.* (1981) respectively). The structure-staining analysis of dyes giving Q-banding patterns was carried out by Curtis and Horobin (1982).

8.1.4 Glycosaminoglycan and mucin stains

Here are discussed two types of dyeing method for identifying glycosaminoglycans; various methods using Alcian Blue; and metachromatic staining.

Alcian Blue — directly and using controlled pH and/or salt content

Objectives

To stain glycosaminoglycan(GA)-rich structures such as cartilage matrix, goblet cell mucin, or mast cell granules; and sometimes to distinguish between sulphated and carboxylated GAGs.

Staining reagents

Alcian Blue is a big, hydrophilic, cationic dye, with a large conjugated system. Its cationic side chains are detached by reaction with nucleophilic groups such as tissue amines or solvent hydroxide ions, yielding an insoluble non-ionic pigment. The alternative dyes, Astra Blue 6BLL and Alcec Blue, have similar properties, but are chemically more stable.

Critical technical steps

First the 'simple' method, then variants.

(1) Stain with aqueous dye solutions — only GAGs stain.
(2) Dehydrate through alcohol, and mount.

Selectivity is controlled by varying the pH and salt content of the dye solution, and the time of staining. The variant giving *general staining of GAGs* has a dyebath adjusted to pH 2.5, contains no extra electrolyte, and uses a short staining time. A varient for distinguishing carboxylated from sulphated GAGs makes use of *controlled pHs*. These are typically over the range pH 0.2–2.5; no extra electrolyte is added, and a short staining time is used. At low pH only sulphated GAGs stain. Another variant makes the carboxylated/sulphated distinction by using *controlled electrolyte concentrations*. When the salt added is magnesium chloride, the concentration range is typically 0.02–0.9*M*, with solutions buffered to pH 5.8; the staining times used are long. At high electrolyte concentrations only sulphated GAGs stain. Alcian Blue shows marked batch variation in purity and solubility, the latter feature exacerbated by the salty solutions. Substitutes for Alcian Blue are not widely available commercially.

Key questions

Why does Alcian Blue stain anionic GAGs, but not anionic DNA and RNA? How do alterations in pH or electrolyte concentration influence the selectivity of staining? Since Alcian Blue is a basic dye, why is it practicable to use alcoholic dehydration?

Mechanism

Staining of the anionic GAGs by Alcian blue is basic dyeing, since the dye is cationic; but selectivity is also influenced by other factors. Consider the *general staining* and the *controlled pH* variants. The non-staining of nuclei and basophilic (that is RNA-rich) cytoplasms is a rate effect. Since Alcian Blue is large and slow diffusing, during short staining times it enters only the more permeable sites; namely swollen, GAG-rich structures (see section 4.3.1). When the pH is varied, sulphated GAGs remain anionic at high acidities, and so stain; whereas carboxylated GAGs loose their anionic character around pH 4–5. The mechanism of the *controlled electrolyte concentration* or 'critical electrolyte concentration' variant, which also distinguishes carboxylated from sulphated GAGs, is more complex. The tendency of sulphated, but not carboxylated, GAGs to stain even at high electrolyte concentrations may be attributed to the high affinity of structures rich in sulphated GAGs. However, this affinity model does not explain some characteristics of the stain. In the absence of rate effects, it is staining of the polyphosphate RNA which is usually the most salt-resistant, and not sulphated GAGs. Moreover many tissue sites do not stain reversibly with Alcian Blue, as the dye reacts with tissue groups.

Small print. Artifactual nuclear staining occurs after prestaining procedures such as enzymic extractions or low pH blockades, which remove components from the tissue, speeding up dye entry (see section 6.2). Alcian Blue's convenient ability to resist extraction into alcoholic dehydration media is due to the dye's large size, large conjugated system, and hydrophilic nature (see Aluminium Haematoxylin, discussed above).

Alternative views. For a more optimistic view of the controlled electrolyte concentration variant (under the heading 'Critical electrolyte concentration method') see a review by Scott (1973). He also argues that failure to stain nucleic acids is due to a sterically induced lack of affinity.

Implications

Rate effects, wide spread of p*K* values, the complexity of histochemical salt effects, and the reactivity of the dye with solvent and tissues make the use of Alcian Blue to determine the anionic character (that is sulphated, phosphated, carboxylated) of unknown tissue biopolymers unreliable. Note also that use of extraction controls will often give rise to positive artifacts.

References

Extensive investigations have been carried out on rate effects in Alcian Blue staining and on the complex effects of electrolytes on staining with cationic dyes (for instance, Goldstein and Horobin, 1974; Horobin and Goldstein, 1974; Bennion and Horobin, 1974).

Staining with metachromatic dyes

Objectives

To stain anionic glycosaminoglycans (GAGs) one colour, and other basophilic sites a contrasting one, using a single dye.

Staining reagents

Dyes giving such metachromatic staining include Azurs A and B, Methylene Blue, Thionin and Toluidine Blue. These are all small, blue, slightly hydrophilic unicationic dyes, with planar conjugated systems. Some of these (for example, Azur B, Methylene Blue and Thionin) are commercially available in pure form, others are inconsistent in composition.

Critical technical steps

A simple outline precedes some variants.

(1) Optional oxidation — enhances, but is not necessary for, metachromasia.
(2) Stain with one of the above dyes in a slightly acidic, aqueous solution — after which anionic GAGs are coloured a metachromatic red/purple, whilst other basophilic materials are an orthochromatic blue.
(3) Optional mordanting — after which dyes resist extraction into alcoholic dehydrating fluids.
(4) Dehydrate and mount.

Prestaining oxidation is usually with permanganate. Over-oxidation can give rise to metachromatic staining of additional, non-GAG materials such as glycogen. The staining step can be made selective for sulphated GAGs by lowering the pH of the dyebath below pH 3. 'Mordanting' refers to a quick immersion in an aqueous uranyl nitrate solution. Stained sections are dried in various ways, such as by blotting followed by air drying, or by dehydration through acetone — alcohol dehydration is avoided unless there is a prior mordant treatment. Mounting media vary; metachromasia is most pronounced when aqueous media are used.

Key questions

What is the cause of metachromasia, and why are all basic dyes not metachromatic? Why do only GAGs stain metachromatically? How does the pre-staining oxidation favour metachromatic staining? What motivates the wide variety of post-staining possibilities?

Mechanism

Staining is basic dyeing, as the dyes are all cationic, and the sites stained are anionic GAGs or nucleic acids. Selectivity results from metachromatic staining being restricted to GAGs. Metachromasia is due to aggregration of dye ions, the interactions between closely adjacent ions giving spectral shifts. Factors facilitating aggregation thus enhance metachromasia. For instance, aggregation places dye cations in close proximity; but flexible anionic GAGs favour charge neutralization and hence, indirectly, metachromasia. Nucleic acids, being intimately bound to bulky proteins, are less flexible. The varied post-staining options reflect the importance of two requirements for maximizing metachromasia in the mounted specimen. First is the need to maintain a high dye concentration in the tissues, since loss of dye can convert metachromatic staining to orthochromatic. As the dyes are small and alcohol soluble this explains such maneouvres as avoiding solvent dehydration altogether, or forming poorly soluble uranyl complexes prior to alcohol dehydration. Second is the need to limit losses of water from the stained tissue, as metachromasia requires the presence of water. GAGs are highly hydrated biopolymers.

Small print. The pre-staining oxidation enhances basophilia by generating extra anionic sites, such as carboxylates from the sugar residues of polysaccharides and GAGs. Not every basic dye gives histochemically useful metachromatic staining. Dye aggregation is facilitated by a low electric charge, a planar conjugated system, and the absence of obstructing bulky substituents — all features found in the dyes listed above, but not universal. Moreover even if dyes do give substantial metachromatic spectral shifts, they will not necessarily be technically useful. The blue dyes used here exhibit perceptually distinct red-purple metachromasia; but the red dye Pyronin Y gives yellow-orange metachromasia, which is insufficiently distinct to the eye for useful staining.

Implications

So many technical factors are involved that standardization would be difficult. If attempted, it should use dyes which are commercially available in pure form; give intense and stable metachromasia; and utilize standardized mounting procedures.

References

The histochemical use of metachromasia has been reviewed, for instance by Pearse (1968).

8.1.5 Lipid and myelin stains

Two different dyeing strategies are considered here. Firstly the use of non-ionic dyes, which stain most classes of lipids. Secondly the use of hydrophobic salts of anionic dyes, to yield selective staining of myelin.

Dyeing of lipids with Sudan Black B and Oil Red O

This vignette discusses the Oil Red and Sudan Black methods as variants of a common procedure.

Objectives

To stain both hydrophilic and hydrophobic lipids; or, with some slight modification, to distinguish between these classes.

Staining reagents

The azo dyes Oil Red and Sudan Black are non-ionic, and are both hydrophobic.

Critical technical steps
(1) Mount cryostat sections onto slides.
(2) Optional bromination — enhances staining of all lipids, but most dramaticcally of lecithin, fatty acids and cholesterol.
(3) Optional extraction with cold anhydrous acetone — to give selective staining of the unextracted hydrophilic lipids.
(4) Stain with azo dye, in partially aqeuous solution — giving strong staining of lipid, plus a paler general background colouration.
(5) Differentiate in an aqueous alcholic solvent, along with a stained de-lipidized control, until the control is colourless.

The dyes are applied from different solvent systems. Sudan Black is often applied as a saturated solution in aqueous ethanol, Oil Red from aqueous isopropanol. The subsequent differentiation steps use aqueous ethanolic and isopropanolic solutions respectively.

Key questions
Why do only lipids stain with these dyes? What is the purpose of having mixed water–alcohol solvents for staining? How does bromination enhance staining, and in particular why are solid or crystalline lipids not strongly stained without the bromination step?

Mechanism
The staining is a partitioning process. In thermodynamic terms, this is driven by the increase in entropy occurring when dye is present in both dyebath and tissue phases, rather than being limited to the dyebath (see section 3.4.2). It is not surprising that a structure-staining correlation based on partition coefficients predicts effective lipid staining by non-ionic dyes; for effective lipid stains log $P>+5$ (see Appendix 1). However, selectivity depends on the non-lipid materials not staining, in spite of the reagent entropy effect. In this context the significant feature of non-lipids — whether glycosaminoglycans, nucleic acids or proteins — is that they are polar, and hence highly hydrated. For the hydrophobic lipid dyes to be inserted into such 'wet' environments would be entropically, and perhaps enthalpically, unfavourable (see section 4.2.2). In this context consider the multiple advantages of the mixed aqueous -alcoholic solvents used in the dyebaths. Such solvents ensure that non-lipids are well hydrated. They also aid partition, since the more hydrophilic the solvent the more hydrophobic dye will pass into the hydrophobic lipid. Finally, extraction losses of hydrophobic lipids are reduced if the dyebath is hydrophilic.

Small print. The Step (2) bromination converts unsaturated lipids — whether cholesteryl esters, fatty acids or phospholipids — into more hydrophobic bromo derivatives. This reduces losses of such compounds into the dyebath; and increases the solubility of the hydrophobic dyes in the lipids. Moreover lipids such as cholesterol which are present in crystalline form stain weakly. The bromination step reduces the melting point of such lipids, enhancing staining. More effective staining of liquid lipids is congruent with a partition mechanism. Both rate and affinity factors are involved, since dye dissolves both faster and in greater quantity in the liquid.

Implications
Since lipid staining is based on partition, and as different lipids are so chemically varied, it is unlikely that all lipid classes will be stained with equal intensity, a differential which the bromination step does something to reduce. Note that extraction controls using chloroform-methanol may be used.

References
An excellent background account of these procedures is available (Bayliss High, 1982). The structure-staining correlation model, and its practical applications, has been discussed (Horobin, 1981; Proctor *et al.*, 1984).

Dyeing of myelin with Luxol Fast Blue
Objectives
To stain myelin, without colouring other lipids.

Staining reagents
Luxol Fast Blue A, G, and M are three unusual acid dyes. The oddity is not a feature of the coloured anions (which are big and hydrophilic, with large conjugated systems) but of the counterions. Instead of the usual hydrophilic sodium ions, these are large hydrophobic organic cations. In this form the dyes are insoluble in water and hydrocarbons, but soluble in polar organic solvents.

Critical technical steps
(1) Fix tissue in formalin, embed in paraffin wax, and mount sections on slides.
(2) Stain overnight with Luxol Fast Blue from hot, slightly acidic, 95% ethanol solution — giving blue colouration of myelin, with other lipids unstained; background staining of acidophilic materials occurs.
(3) Differentiate alternatively in aqueous alkali and 70% aqueous ethanol — most background staining is removed, though elastic fibres, eosinophil granules and erythrocytes remain coloured.

The Luxol method is technically simpler and more flexible than traditional Weigert–Pal Haematoxylin procedures. If Luxol dyes are not available, several acid dyes may be substituted, after precipitating them from aqueous solution in the form of their ditolyguanidinium salts.

Key questions
Why does myelin stain, but not hydrophobic lipids? Why should acid dyes stain lipids, but not cytoplasms, collagen fibres and so on. Why is the dyebath alcoholic, and the differentiator aqueous alkali?

Mechanism
The presence of the hydrophobic cations can facilitate staining of myelin in two ways. First, because in non-aqueous solvents with low dielectric constants, cation–anion pairing is enhanced. The use of large organic counterions results in hydrophobic ion pairs being formed, which partition into myelin in which hydrophilic lipids are

plentiful. In keeping with this partition model, use of more hydrophobic solvents reduces myelin staining. Second, because hydrophobic cations may adsorb onto the myelin, they give rise to a cationized surface to absorb dye anions. Dewaxed paraffin sections contain few hydrophobic structures other than phospholipids. Elastin is a hydrophobic protein, and is stained with the Luxols. However, as the dyes are large conjugated systems the staining may be due to van der Waals attractions (see section 8.1.2, stains for elastic fibres). The background staining occurring in Step (2) is acid dyeing of cationic tissue proteins by the dye anions. In the subsequent alkaline differentiation, proteins become anionic, and dye anions are lost from most structures. The exceptions — elastic fibres, eosinophil granules, erythrocytes — are all dense, and so slow to de-stain. Such a rate control mechanism is aided by the large size of the Luxol dyes. Moreover elastic fibres have considerable non-coulombic affinity; and eosinophil granules contain highly basic proteins, and will still be cationic under alkaline conditions. As the differentiating solutions are largely aqueous, extraction of hydrophobic dye ion-pairs from the myelin will be limited; moreover myelin is itself dense material.

Alternative views. The stain was originally conceived as specific for phospholipids, in a chemical sense (Pearse, 1960).

Implications

Selectivity is dependent on differentiation, so strong staining of fine myelinated fibres may be problematic. This procedure is not limited by the commercial availability of the Luxol dyes, nor is it limited to blue staining, since a number of widely available acid dyes can be used after their precipitation as ditotylguanidinium salts.

References

For a discussion of the role of hydrophobic cations in Luxol dyeing of myelin, and the preparation of further 'Luxol type' dyes of various colours, see Clasen *et al.* (1973).

8.1.6 Nucleic acid stains

These dyeing methods demonstrate polymeric nucleic acids, specifically the deoxyribonucleic acid of chromatin and the ribonucleic acid of nucleoli and ribosomes: subsequently termed DNA and RNA respectively. The first vignette describes techniques which stain both DNA and RNA, without distinguishing between them. The second a method which both stains and distinguishes DNA and RNA. The final vignette is of a technique which stains only RNA. The standard method for selectively staining DNA is the Feulgen nucleal procedure (see section 8.3.1), which does not involve dyeing.

Nucleic acid dyeing with Cresyl Fast Violet and Gallocyanin Chrome Alum

These two methods are treated as variants of a single procedure.

Objectives

To demonstrate nucleic acid-rich sites. With the dyes discussed, selective staining of DNA and RNA require pre-staining extraction of one of the nucleic acids.

Staining reagents
Cresyl Fast Violet (CFV) is a fairly small, slightly hydrophilic, cationic dye. Gallocyanin Chrome Alum (GCA) is a big, hydrophilic metal complex dye, with a large conjugated system. GCA solutions are acidic and, in addition to the blue metal complex dye, contain the red dye Gallocyanin plus chromium(III) cations.

Critical technical steps
A general outline is given, followed by details specific to the use of CFV and GCA.

(1) Fix tissue, process through paraffin, mount on slides.
(2) Optional enzymic extraction with deoxyribonuclease or ribonuclease — this converts this general method into one showing some selectivity for RNA and DNA respectively.
(3) Treat with aqueous acidic dye solutions — material staining includes basophilic cytoplasms, chromatin, and nucleoli; also cartilage matrix and mast cell granules.
(4) Alcoholic differentiation (CFV only).
(5) Dehydrate through the alcohols, and mount.

The features specific to the CFV and GCA *variants* will now be specified. The mode of fixation is without marked effect with GCA; with CFV, alcohol or Carnoy's fluid are sometimes found to be superior to formalin. CFV staining is quite rapid, typically 30 min or less; and usually requires alcoholic differentiation. GCA is very slow staining, needing hours or even days; however, no differentiation is called for. GCA, unlike CFV, can be dehydrated at leisure through the alcohols.

Key questions
Why are nucleic acids stained selectively? Can the reasons for the staining of other materials be specified? Why is GCA a progressive stain, whilst CFV requires differentiation? Why can sections stained with GCA, unlike CFV, be dehydrated through the alcohols?

Mechanism
The selective staining is basic dyeing: the cationic dyes bind to anionic nucleic acids, and certain anionic glycosaminoglycans, but not to proteins, which at the low pH of the dyebaths are cationic (see section 3.3.1). The fast staining of CFV may be attributed to the dye's small size and small conjugated system, GCA being twice as large, and more than twice as slow staining (see section 4.3.1). As a consequence of the rapid CFV staining, thick sections (for instance Nissl stained neuropathological material) require differentiation, to remove excess stain from more superficial layers of the section.

Small print. The large conjugated system of GCA, with consequent enhancement of non-coulombic binding due to van der Waals attractions, accounts for insensitivity to mode of fixation. The smaller CFV will be more influenced by charge and hydrophobicity of the tissue, and hence to fixative action. Enzymic extraction of nucleic acids is influenced by fixation, though this is not made clear in some practical texts. Poorly fixed tissue loses RNA non-specifically into aqueous solutions; whilst well-fixed tissue resists extraction by either enzyme. The resistance of GCA to extraction into dehydrating alcohols is due to its hydrophilicity and large conjugated size, typical of metal complex dyes (see section 4.3.1 and 6.4). Neither factor holds for CFV, so alcohol dehydration must proceed with caution, as is the case for most commonly used basic dyes. Occasional red staining of cartilage matrix is due to unusually high Gallocyanin concentrations in the GCA mixture; GCA itself stains cartilage and mast cell granules blue-black.

Alternative views. GCA has been regarded as a mordant dye, binding to nucleic acid phosphate groups via the chromium ion (Sandritter *et al.*, 1966).

Implications

Since GCA is innately a mixture, standardization of this reagent would not be easy. In spite of this the dye has been successfully used for the quantitative estimation of nucleic acids. Enzyme extraction controls may be used with these dyes.

References

The structure and staining mechanism of GCA, together with comments on the composition of GCA solutions, have been discussed by Marshall and Horobin (1972). The use of GCA for quantitative histochemistry is described by Sandritter *et al.* (1966). A discussion of differences in the staining mechanisms of the usual basic dyes and of cationic metal complex dyes is found in another paper of Marshall and Horobin's (1973a).

Distinguishing DNA from RNA by Methyl Green–Pyronin staining

Objectives

To stain DNA and RNA green and red respectively

Staining reagents

Methyl Green is cationic, hydrophilic, of moderate size both overall and in terms of its conjugated system. This dye often contains traces of a decomposition product, Crystal Violet. Many commercial samples labelled Methyl Green are actually Ethyl Green; fortunately the staining produced by these two dyes is indistinguishable. The red dye Pyronin Y is cationic, only slightly hydrophilic, and small. Commercial Pyronin Y is commonly impure. Other pink dyes may be substituted, and even when the dye in the bottle is Pyronin Y, it is often grossly diluted with colourless materials. These problems are, in principle, avoidable since both dyes are commercially available in pure form.

Critical technical steps

There is a wide range of Methyl Green–Pyronin (MGP) variants; some of which are mentioned below, following the outline method.

(1) Fix (most agents can be used), embed in paraffin wax, mount sections on slides.
(2) Stain in an aqueous, slightly acidic, dyebath containing both Methyl Green and Pyronin — which gives green to purple nuclei, red nucleoli and basophilic cytoplasms, plus orange background staining of glycosaminoglycan-rich structures.
(3) Rinse briefly in water.

(4) Blot dry, dehydrate through some organic solvent, and mount.

Each step has variants. Though most fixatives can be used, the final colour balance is fixative-dependent. Thus fixation in Carnoy's fluid favours green staining, whereas Zenker's favours red. Analogously acid decalcification favours green colouration. Such influences are compensated by alterations in relative and absolute dye concentrations. Staining has a number of variations. These include, in addition to dye concentrations just mentioned, the staining time and temperature, the pH, and the nature and amount of colourless co-solutes present in the dyebath. Post-staining steps show similar variety. Rinses differ in length and temperature: some are extremely short, and some extremely cold. Different dehydrating regimes are used in the various protocols, for example using acetone-xylene, ethanol-butanol, or *t*-butanol. Faced with this plethora of options, it is a relief to know that if pure dyes are used, widely applicable standardized procedures may be specified.

Key questions
Why does one cationic dye stain DNA and another RNA? Why are post-staining manipulations so technically tricky? Indeed why does MGP staining have so *many* variants? Why is the background staining orange?

Mechanism
This is basic dyeing, as the cationic dyes bind to anionic nucleic acids and glycosaminoglycans (see section 3.3.1). The basis of selective staining of DNA and RNA is not entirely clear. The mechanism involves dye competition, since either dye applied individually stains both nucleic acids. One possible cause of selectivity is rate control. Methyl Green is larger than Pyronin, and ribosome-rich cytoplasms stain more slowly than does chromatin; so a simple rate model does predict the observed staining pattern (see sections 4.3.2 and 4.3.1). Such a model is congruent with the effects of fixation and acid decalcification noted previously (see sections 6.1 and 6.2). However, a recent study found the green–red staining pattern was maintained even after extremely prolonged staining times, a puzzle from the rate control viewpoint.

Small print. The variety of post-staining manipulations largely reflects efforts to limit or prevent loss of Pyronin. Consistent with this dye being the most readily extracted is its small size, which facilitates diffusion; and its more hydrophobic nature, which enhances alcohol solubility. The single most disruptive technical factor is variation in the Pyronin samples. Orange colouration of cartilage matrix, goblet cell mucin and mast cell granules is metachromatic staining by Pyronin (see section 8.1.4).

Alternative views. Several workers have proposed that differential MGP staining is affinity-controlled. For instance the significance of salt effects has been emphasized by Scott (1967).

Implications
Pure dyes are required for standardized MGP methods.

References
The said methods have been described by Lyon and co-workers (Hoyer *et al.*, 1986). A rate control mechanism for MGP was described by Goldstein (1961).

Staining RNA with Cuprolinic Blue

Objectives
To stain RNA, whilst leaving DNA unstained.

Staining reagent
Cuprolinic Blue is big and hydrophilic; its four cationic groups are symmetrically placed within the dye's large and rigid conjugated system.

Critical technical steps
(1) Stain with salty, slightly acidic aqueous Cuprolinic Blue solution — which only stains RNA.
(2) Dehydrate through the alcohols, and mount.

With the aid of external standards, this procedure can be used for the quantitative histochemistry of RNA.

Key questions
Why does Cuprolinic Blue stain RNA, but not DNA, or indeed other polyanions? Why is alcoholic dehydration possible with this basic dye?

Mechanism
Staining is basic dyeing, with cationic dye binding to anionic RNA; under dyebath conditions proteins will be cationic (see section 3.3.1). Selectivity for RNA is a result of adding electrolyte. Although histochemical salt effects are complex, a mechanism can be proposed. Addition of salt to a dyebath typically reduces cationic dye–anionic polymer binding, doubtless the reason why all tissue anions stain unless the electrolyte is added. The question now arises: why is staining of RNA the most salt-resistant? If simple ion-exchange systems are taken as models, then for a hydrophilic, non-aggregating dye such as Cuprolinic Blue the probable reason for this selectivity pattern is the impermeable nature of the ribosomes which contain the RNA. Note that both aggregating and hydrophobic dyes — for instance Alcian Blue and Crystal Violet respectively — show different patterns of salt effects.

Small print. The reason preparations stained with Cuprolinic Blue may be dehydrated through alcohols is the large size and hydrophilic character of the dye (see Aluminium Haematoxylin (section 8.1.1, Haematoxylin and Eosin)).

Alternative views. The originator of this use of Cuprolinic Blue regarded its selectivity as being due to specific ion effects (Scott, 1972).

References
The use of the Cuprolinic Blue method for quantitative histochemistry has been described (Mendelson *et al.*, 1983; Tas *et al.*, 1983). Salt inhibition of basic dyeing has been discussed many times, for example from the viewpoint of the Donnan equilibrium (Bennion and Horobin, 1974), and the critical electrolyte concentration theory (Scott, 1973). The role of swelling and crosslinking in selectivity is discussed, in the context of ion-exchange systems, by Marcus and Kertes (1969). For examples of the different salt effects seen with aggregating and hydrophobic dyes see Scott (1972) and Horobin and Goldstein (1974).

8.1.7 Polysaccharide stains

These vignettes discuss three methods: the Chorantine Fast Green procedure for the plant polysaccharide callose; Best's carmine procedure for staining the animal storage polysaccharide glycogen; and the traditional 'iodine' stain for the plant polysaccharide starch. The latter reagent is of course not a dye; however, the method fits as well here as anywhere. Staining of an additional polysaccharide-containing substance, anionic glycosaminoglycans, is discussed in section 8.1.4. Chitin can be stained by the periodic acid–Schiff procedure, discussed in section 8.3.4.

Staining callose with Chlorantine Fast Green

Objectives

To stain callose, but not cellulose or other biopolymers.

Staining reagents

Chlorantine Fast Green is an anionic, hydrophilic dye, with a large conjugated system and a free hydrogen bonding group: as are substitutes such as Evans Blue and Sirius Rose BB.

Critical technical steps

(1) Stain briefly at room temperature in an aqueous dyebath — which gives colouration of callose and cell cytoplasms.
(2) Wash well in water — after which only callose retains dye.

Methods substituting the dyes named above give analogous results.

Key questions

Why does callose stain, whilst cellulose, and other common plant biopolymers, do not?

Mechanism

Dye binding to callose involves van der Waals attractions plus hydrogen bonding. Cellulose, unique among polysaccharides in having few free hydroxyl groups able to form *inter*molecular hydrogen bonds, fails to stain. Factors aiding selectivity are the anionic and hydrophilic character of the dye: so dye does not bind to anionic hydrophobic lignins of xylem vessels, nor does hydrophobic bonding assist binding to proteins or suberin.

Implications

Given the range of substitutes available, the method is not restricted to green colours, nor by the limited commercial availability of the title dye. If the mechanism proposed is correct, other polysaccharides with free hydrogen bonding groups may also stain.

Reference

The technical aspects are described by Pearce (1986).

Staining glycogen with Best's carmine

Objectives

To stain the polysaccharide glycogen, without staining other components of mammalian cells.

Staining reagents

The active constituent of Carmine is Carminic Acid: a red anionic dye of moderate size and hydrophilicity, one of whose aromatic hydroxyl groups ionizes under alkaline conditions giving an aryloxy anion free to form intermolecular hydrogen bonds. Carmine contains variable amounts of protein, and of aluminium, calcium and ferric ions as well as the dye. Substitutes — such as Alizarin Brilliant Blue BS, Alizarin Red S, Gallein or Haematoxylin — all have small conjugated systems, and under alkaline conditions aryloxy anions free to form hydrogen bonds.

Critical technical steps

(1) Stain with Carminic Acid dissolved in an alkaline, salty aqueous methanolic solution — giving staining of glycogen.
(2) Wash with aqueous alcohol.

The staining solution is made alkaline by ammonia plus potassium carbonate; with extra salt added in the form of potassium chloride. Exposure of stained sections to water rapidly extracts all Carminic Acid. *Variant* procedures, which stain with the substitute dyes named above, give virtually identical staining patterns.

Key questions

Why does glycogen, and little else, stain with Carminic Acid? Why is the dyebath formulation so complex? Why is water washing of stained slides avoided? When a nuclear counterstain is not used, the nuclei are sometimes stained red — why?

Mechanism

Glycogen staining involves hydrogen bonding between an anionic aryloxy group on the dye and a hydroxyl group on the glycogen. Hence the need for a pH high enough to substantially ionize the aromatic hydroxyl. Hence also the need for a largely non-aqueous dyebath and post-staining wash solvent; otherwise water–dye and water–glycogen hydrogen bonding compete with dye–glycogen links. It is possible that hydration of the many inorganic ions present also significantly reduces the amount of free water. Staining selectivity is partly controlled by the plentiful free hydroxyl groups on glycogen, no other biopolymer found in the animal cell carries so many hydrogen donors. Carminic Acid is anionic, but background acid dyeing does not occur since tissue proteins are themselves anionic under alkaline conditions.

Small print. Occasional nuclear staining is probably due to traces of cationic metal complex dyes, formed by reaction of Carminic Acid with contaminating aluminium and ferric cations. Non-coulombic background staining is minor, as the dye is hydrophilic and its conjugated system only of moderate size; neither hydrophobic bonding nor van der Waals attractions will be significant.

Implications
This procedure is simple and cheap — but not specific for glycogen. Consistent with the above mechanism, biopolymers carrying large numbers of free hydrogen donors — for instance cellulose, neutral GAGs and starch — also stain with Best's carmine. Given the range of substitutes, the method is not restricted to red colouration nor to using a single natural product. Should standardization be attempted, certain simplifications could be investigated: for example replacing volatile ammonia with another alkali, and the use of pure Carminic Acid, or some other dye less variable in composition than Carmine.

References
Discussions of the technical factors, and of the hydrogen bonding mechanism, are provided by Cook (1982) and Horobin and Murgatroyd (1971) respectively.

Staining starch with iodine
Objective
To stain starch blue, in contrast with other polysaccharides.

Staining reagent
This is produced by mixing iodine with potassium iodide; the active constituent is the large hydrophobic, highly polarizable, triiodide anion.

Critical technical steps
(1) Stain with an aqueous 'iodine' solution — giving blue colouration of starch, with cellulose and glycogen being red-brown.
(2) Mount in water.

These preparations are not permanent.

Key questions
Why does only starch stain blue? Why is the method not permanent? If this procedure is not dyeing, what is it?

Mechanism
The triiodide ion forms an intensely blue clathrate with amylose, the linear polysaccharide component of starch. The branched amylopectin component, and also other polysaccharides such as cellulose and glycogen, also form clathrates. However, these are pale brown in colour. The blue complex arises due to hydrophobic bonding. In an aqueous milieu, the amylose chains adopt the form of α-helices; the central channels of which are relatively hydrophobic, and enclose the hydrophobic triiodide ions. Consequently non-aqueous mountants cannot be used. The triiodide ion probably binds weakly to the amylose by van der Waals forces. However, even in aqueous media the triiodide anion is in equilibrium with iodine plus iodide; and the volatile element iodine is slowly lost.

Small print. Although the triiodide anion is certainly not a dye, it is noteworthy that the factors involved in its uptake are just those involved in many dyeing procedures.

Reference
An account of the formation of iodine/starch complexes is given by Whistler and Daniel (1979).

8.2 SILVER STAINING AND RELATED METAL IMPREGNATION

Metal impregnations use silver salts, or chemically equivalent compounds, as reagents. Colouration is usually due to deposition of microcrystals of a metal, although in certain methods metal salts form the final stain. The affinity for the tissue of this final stain is largely due to stain–stain attractions, with tissue–stain bonding being trivial (see section 3.2.1). Formation of microcrystals often involves one or more catalytic steps. Even small amounts of a tissue component can produce an intense colouration, a staining amplification analogous to that achieved in enzyme histochemistry (see section 3.5.4). Detection of the microcrystals by reflected light microscopy is usually possible, further enhancing sensitivity. Amplification can involve geometry as well as intensity, since the microcrystals can be much larger than the tissue site triggering their formation. The corollary is a loss of spatial resolution.

Silver staining, and related impregnation methods, can be used to demonstate an amazing range of tissue components, from a particular polysaccharide, to the nucleolar organizing regions of chromosomes, to specific cell types in the central nervous system. These impregnation methods have much in common, both technically and chemically. To facilitate a rational account of silver staining, certain specialized chemical information is introduced in section 8.2.1. Subsequently vignettes illustrating a range of metal impregnation methods are given.

8.2.1 Underlying processes of silver staining

The crucial event of silver staining methods is reduction of silver to its metallic form. For clarity this will be written as the transformation of silver(I) to silver(0). Usually this corresponds to silver cations being converted to electrically neutral atoms. In the simplest case this is due to a reducing agent in the tissue producing sufficient metallic silver to directly visualize the tissue structure. The reducing agents generating such ‘argentaffin staining’ may be native or induced. Proteins of APUD cells are innately rich in thiol and phenolic groups, whilst polysaccharides such as glycogen and fungal chitin can be oxidized with chromate, periodate or permanganate to yield aldehyde groups. The silver(I) is applied in a variety of forms, for example as hydrated cations, or coordination complexes with ammonia or pyridine.

Tissue reducing reagents may give rise to tiny aggregates of silver atoms, whilst failing to give metallic crystals visible in the microscope. This is analogous to the formation of a latent image in a photographic emulsion. Such ‘latent images’ can be detected by their ability to catalyse the reduction of further silver(I) on their surfaces. In this second reduction step, solutions of formaldehyde, hydroquinone, pyrogallol, sulphite ion or other reducing agents are applied to the specimen. The additional silver(I) involved in these secondary reductions may also be directly acquired from an applied solution. A solution containing both silver(I) and a reducing agent is termed a physical developer, which must also contain an inhibiter (such as gelatin or dextrin) to stop silver being reduced in, and precipitated from, the solution. However, the silver(I) which is reduced at latent image sites may, instead, come from the prior reversible uptake of silver(I) by proteins or other tissue sites. Thus silver(I) can be coordinatively bound by imidazole residues, or by phosphate groups of DNA, or the hydroxyapatite in mineralized bone. Such silver(I) is mobilized during application of the reducing agent, and some of the metal ions diffuse to the catalytic sites.

When such a redistribution of silver occurs, the final staining image can be influenced by various rate effects. If the rate of diffusion of silver(I) is high compared to the rate of reduction, staining will be weak since much silver(I) will be lost into the reducing bath. If diffusion is slow compared to the rate of reduction, positive artifacts may occur due to non-catalytic formation of metallic silver, especially in dense structures such as red blood cells and nucleoli which reduce diffusion rates (see section 4.3.1). Staining rates may also be influenced by light, since photoreduction of reversibly absorbed silver(I) can occur.

Catalytic reduction of silver(I) is induced by the presence of various ionic crystals, as well as by metallic microcrystals and the latent images discussed above. If low concentrations of metal ions present in live cells are converted into sulphide precipitates, during or before fixation, the resulting deposits can catalyse formation of metallic silver from physical developers. Analogously, metal ions (for instance mercuric, phosphomolybdate, uranyl) may be selectively deposited in initial mordanting or priming steps, and the resulting bound metals can then act as catalytic sites. When metal-containing fixative agents (for example mercury salts, or osmium tetroxide) are used, this phenomenon can occur artifactually, giving rise to non-specific background staining. Some stains have catalytic properties too, which can be used to generate an amplifying secondary metal staining. An example is provided by the polymeric reaction products arising during enzyme histochemical demonstration of peroxidase. These catalyse reduction of silver(I), an event currently exploited to increase the sensitivity of peroxidase-labelled antibodies.

After each procedural step it is important that unreacted reagents are removed. This is particularly so for unreduced silver(I) bound to tissue proteins; hence the washing of specimens with sodium thiosulphate. This forms the water soluble complex ion $Ag(S_2O_3)^{3-}$, which is readily extracted into the aqueous solution. Another general point is that many silver stains involve coordination of the metal ion by ammonia or pyridine, and so have alkaline solutions. The resulting tissue swelling may assist reagent entry, which is especially important with the thick sections and tissue blocks often used in neuroanatomical work. Another feature aiding reagent penetration of such material is the fact that pyridine is a lipid solvent.

As this account has shown, cleanliness, the concentration of silver and of reducing agent, fixation, light intensity, pH, specimen thickness, temperature, times of reaction, water quality, and washing can all influence staining — sometimes at more than one point in the procedure. Difficulties experienced in achieving reproducibility of many silver methods is therefore not surprising.

References

Methodological papers and empirical observations abound, as reviewed for instance by Velikan and Velikan (1972). Critical experimentation and theory building is less common. The above account leans heavily on the work of Gallyas (1979) and Peters

(1955). For convenience of reference, technical details of all the methods discussed in the following vignettes are described — and indexed — in a standard procedures book (Bancroft and Stevens, 1982).

8.2.2 Direct reduction procedures

Both the procedures discussed as examples involve non-catalytic reduction of silver(I) to metallic silver crystals; one method utilizes photoreduction the other does not.

Singh's stain for argentaffin cells

Objectives

To demonstrate certain cells, for instance those of the APUD series, by staining their secretion granules.

Staining reagents

Silver(I) is present as silver ammine cations, $Ag(NH_3)_2^{1+}$.

Critical technical steps

(1) Immerse slides in warm, aqueous silver(I) solution, until sections appear light brown — giving black staining of secretory granules.
(2) Treat with ('fix in') aqueous sodium thiosulphate.

Key questions

What is special about the secretion granules? Why is a post-staining fixation step needed?

Mechanism

The granules of argentaffin cells contain high concentrations of compounds such as 5-hydroxytryptamine. These directly reduce silver(I), yielding microcrystals of metallic silver. Staining is monitored, and halted as soon as brown colouration is visible, eliminating background staining due to the slower reduction of silver(I) bound to other sites. Since reduction in such sites could eventually occur in mounted sections, unreduced silver is removed by a thiosulphate wash: in this context the term 'fix' derives from photography, not histoprocessing.

Implications

This technically simple procedure does require sections to be individually monitored to obtain selective staining. The method is specific neither for 5-hydroxytryptamine nor APUD cells; metallic silver will be deposited wherever suitable reducing agents exist in high enough concentration.

Von Kossa's stain for minerialized bone

Objective

To stain mineralized bone, without staining osteoid or other material.

Staining reagent

This is the hydrated silver(I) cation.

Critical technical steps

(1) Stain in brightly illuminated aqueous silver nitrate until bone darkens — giving black mineralized bone, with other structures being colourless.
(2) Wash in aqueous sodium thiosulphate solution.

Osteoid and other tissue components may be counterstained by any suitable method. For instance van Gieson's stain (see section 8.1.1) colours osteoid red, and cell cytoplasms yellowish.

Key questions

Is illumination necessary? Why is mineralized bone stained, but not the adjacent osteoid? Can *any* counterstain be applied?

Mechanism

Silver(I) ions are adsorbed by the proteins of the tissue and, especially, the numerous phosphate groups of the hydroxyapatite of the bone. Photoreduction occurs most rapidly in the hydroxyapatite; the reaction is monitored to prevent overstaining giving rise to background staining. Unwanted non-selective staining is further limited by removal of unreduced silver(I) by a thiosulphate wash. The chemical stability of metallic silver permits most counterstains to be applied, without risk of removing the silver stain.

Implications

This technically simple method does require sections to be individually monitored to ensure selectivity. Moreover staining does vary with specimen thickness, since the initial, superficial deposition of metallic silver limits illumination of the deeper layers of bone.

8.2.3 Catalytic reduction procedures

Each of the methods discussed involves catalytic reduction steps. Two procedures are catalysed by latent images produced during prior silver impregnation, one by non-silver metal ions deposited during an initial priming procedure. Two of the methods also utilize another catalytic process, gold toning. In one procedure the catalytically reduced silver(I) comes from a secondary developer; in another from silver(I) previously absorbed onto the tissue. A further stain utilizing the catalytic reduction of silver(I), namely an intensification method for immunostaining, is discussed in section 8.5.

Bielschowsky's stain for axons

Objectives

To stain axons, but not nerve cell bodies or non-neuronal material.

Staining reagents

In this complicated method silver(I) is applied in three forms. First as the hydrated cation; then as the cationic silver ammine; and finally as a component of a secondary developer, containing silver(I), formaldehyde, citrate, and a little acid. The active constituent of the gold chloride used for the toning proceduring is the $AuCl_4^-$.

Critical technical steps

(1) Incubate in the dark with a warm aqueous silver nitrate solution.
(2) Incubate in warm aqueous silver ammine solution.
(3) Incubate in a physical developer — giving black staining of axons.
(4) Tone briefly in gold chloride solution — giving an intensified stain.
(5) Wash in aqueous sodium thiosulphate solution.

Key questions

Why are there multiple applications of silver(I)? What is involved in gold toning?

Mechanism

During the initial incubations, enough silver(I) is reduced in the axons to form latent images. The reasons for the selectivity are uncertain. Once formed these clumps of silver(0) atoms catalyse further reduction of silver(I), yielding visible staining. In this second reduction, both silver(I) and reducing agent come from the physical developer. Precipitation of silver(0) from this solution is inhibited by complexing metal ions by citrate. Deposited silver metal catalyses reduction of gold chloride to the metal. In the absence of such catalysis, gold chloride reacts non-specifically with most tissue proteins. Unreduced metal ions are removed by complexation with thiosulphate anions.

Cajal's astrocyte stain

Objectives

Staining of astrocytes, without staining other fibres or cell types of the central nervous system.

Staining reagents

An aqueous mixture of gold and mercuric chlorides.

Critical technical steps

(1) Incubate sections in the dark with aqueous gold/mercury staining solution for several hours *or* until sections darken — giving an intense and selective staining of astrocytes.
(2) Wash in aqueous sodium thiosulphate solution.

If the incubation temperature is raised, the astrocytes stain almost black.

Key questions

Why does the stain contain two different metal ions? Are they both necessary? What is special about astrocytes?

Mechanism

Gold chloride reacts non-specifically with most proteins, depositing metallic gold, and treatment of sections with a mercury-free staining solution yields only diffuse staining. There is probably a fast reaction of mercuric ion with astrocytes, depositing mercury in the cells, catalysing the reduction of gold chloride. Once this process is initiated, gold chloride will be reduced autocatalytically, depositing metallic gold. The selective binding of mercury ions to astrocytes is probably due to their reaction with the acidic (that is, carboxylate-rich) protein plentiful in astrocyte filaments. Note that slow, non-catalytic, reduction of gold chloride is prevented by the thiosulphate wash, which removes unreduced metal.

Gordon and Sweet's reticulin fibre stain

Objectives

To stain reticulin fibres, that is connective tissue fibres rich in type III collagen, found in liver, lymph nodes and spleen, whilst staining other collagen fibre types in a paler and contrasting colour.

Staining reagent

The silver(I) species is a cationic silver ammine.

Critical technical steps

(1) Oxidize with an aqueous potassium permanganate solution.
(2) Optional treatment with aqueous iron alum solution.
(3) Incubate with the aqueous silver(I) solution.
(4) Treat with an aqueous formaldehyde solution — giving black reticulin fibres; grey or black background staining occurs, for instance in some lysosomes and, if the iron alum step is omitted, cell nuclei; most collagen fibres stain pale brown.
(5) Tone in gold chloride.

Key questions

What is special about reticulin fibres? In the step using formalin as a reducing agent, what is the source of the silver(I)? What are the functions of the pre-silver incubation treatments with permanganate and iron alum?

Mechanism

The pre-incubation oxidation converts thiols and disulphides into sulphonate anions, and generates aldehyde groups from saccharide residues. In the incubation step silver(I) cations will bind to proteins, especially to the more anionic sulphonated forms, and also to other tissue anions, such as DNA and GAGs. During incubation the aldehyde groups of reticulin fibres will reduce sufficient silver(I) to produce silver(0) latent images. Subsequent treatment with formalin results in the catalytic reduction of silver(I), released from proteins, at these sites. The selectivity for reticulin fibres probably reflects their relatively high concentration of saccharide residues, compared to other collagen fibre types. Moreover reticulin fibres are made up of type III collagen whose constituent polypeptide chains contain oxidizable disulphide bridges, absent in the polypeptides of type I and II collagens. Enough silver(I) is bound to lysosomal GAGs and to DNA to give visible deposits of metallic silver following the formaldehyde reduction. Inhibition of the nuclear staining by pre-incubation treatment with iron alum is probably due to the binding of iron and aluminium cations to the anionic DNA competing with subsequent uptake of silver(I) cations at these ionic sites.

8.3 REACTIVE AND CHROMOGENIC PROCEDURES

Although one of the methods discussed dates from 1867, the blooming of such procedures was probably the 1950s—with its emphasis on histo*chemistry* and cyto*chemistry*. Unlike dyeing procedures, these methods involve reactions which can be represented by traditional chemical equations, for example breaking or making covalent bonds, changes in oxidation state, and precipitation of ionic crystals.

However, before such reactions can occur the necessary reagents must be present in the tissues. Consequently, as with dyeing, rate and affinity factors are important. Moreover the final reaction products are sometimes bound to the tissue covalently, but in other cases the stain–tissue affinity is due to those factors familiar from discussion of dyeing. So although reactive and chromogenic methods are sometimes regarded as 'simple coordination/inorganic/organic chemistry' they are better seen as procedures just as subtle as dyeing, since only part of their mechanism can be represented in straightforward chemical symbols. Ignorance of this complexity can lead to unnecessary errors.

8.3.1 A stain for DNA

The Feulgen nucleal method

Objective
To stain DNA, whilst not staining RNA; or indeed other biopolymers.

Staining reagent
Schiff reagent contains several necessary components: a colourless, hydrophilic and cationic derivative of Basic Fuchsin, with a sulphite anion covalently attached to the bridging carbon atom; plus hydrogen and sulphite ions.

Critical technical steps
(1) Fix, avoiding acidic fixative agents such as Bouin's fluid.
(2) Hydrolyse with aqueous hydrochloric acid.
(3) Treat with Schiff reagent.
(4) Wash extensively with water—giving magenta staining of DNA.

To achieve the most intense staining, the hydrolysis time must be adjusted to suit the fixative agent used. Variants: there are two types of substitute for Schiff reagent. These are either basic dyes carrying free amino groups, treated with acidic sulphite to make pseudo-Schiff reagents, or Pararosanilin in acidic alcohol.

Key questions
Why is the staining only of DNA, and not, for instance, of RNA? Why is the method sensitive to fixation?

Mechanism
The final magenta stain is bound to the DNA residue by covalent bonds. However, it is the acid hydrolysis which controls selectivity, since it generates aldehyde groups by depurination of DNA. No other tissue components yield aldehydes during this step. Schiff reagent demonstrates aldehydes non-specifically, by reacting with these

moieties, whatever their origin, to yield colourless alkyl sulphonate compounds. Removal of the sulphite ion from the bridging carbons, by the extensive water wash of Step (4), then generates the triphenylmethane chromophore, and hence coloured derivatives. The phosphate esters of the backbone of the DNA polymer are also acid hydrolysed, albeit at a slower rate. Continuing hydrolysis thus increases the number of available aldehydes due to depurination, whilst decreasing the number of aldehydes as whole polynucleotide sequences are separated and diffuse away. The maximum staining intensity attainable thus rises to a maximum, then falls. Fixation influences this competitive process. Firstly, if there was acid present in the fixative agent much DNA may be lost prior to staining. Second, the period of hydrolysis required to give maximum staining intensity is fixative-dependent, probably via influencing the rate at which hydrolysed polynucleotide fragments diffuse from the chromatin.

Small print. Probably the pseudo-Schiff reagents act in a similar way, whilst the use of Pararosanilin in acid alcohol as a substitute aldehyde reagent will involve slightly different covalent links to the DNA residues.

Alternative views. Until the alkyl sulphonate character of the histochemical reaction product was established, there were several competing notions as to the precise organic chemistry of the reaction of Schiff reagent with depurinated DNA, as reviewed by Kasten (1960). Some writers have not yet noticed that this confusion is now clarified.

Implications

As the aldehyde reagent is non-specific, non-hydrolysed controls must be used. The method can be used quantitatively but, owing to the competitive rate effects occurring during hydrolysis, only with specimens of the same geometry and mode of fixation.

References

The alkyl sulphonate nature of the final histochemical reaction product of Schiff reagent with aldehydes was demonstrated by Nettleton and Carpenter (1977). Kasten has clarified many aspects of the Feulgen nucleal method, see his summarizing papers (Kasten, 1960, 1971).

8.3.2 Stains for lipids

The two methods discussed here illustrate something of the rag-bag character of the reactive/chromogenic group of stains. The first procedure involves a colour generating step in which a metal coordination compound is formed, whilst selectivity is controlled by the prior precipitation of a copper salt. In the second method selectivity is controlled by the peculiarities of the organic chemistry of osmium tetroxide, and the subsequent generation of colour results from ill-defined reduction reactions probably involving tissue proteins.

Copper–Rubeanic acid for free fatty acids

Objective

To stain free, that is non-esterified, fatty acids.

Staining reagents
The copper is applied in the form of copper(II) acetate. Rubeanic acid (that is dithiooxamide) is a small, non-ionic compound which reacts with copper(II) ions to give a dark green chelate. Ethylenediaminetetracetic acid (EDTA) is a non-specific metal chelating compound.

Critical technical steps
(1) Fix tissue in calcium-formal; or post-fix cryostat sections in the same.
(2) Treat sections with 1 M aqueous hydrochloric acid, at room temperature, for 30 min.
(3) As a control, extract a section with chilled, dry acetone to remove fatty acids; process the control in parallel with non-extracted sections.
(4) Immerse in an aqueous solution of copper(II) acetate for an extended period.
(5) Wash very briefly in aqueous EDTA.
(6) Immerse in Rubeanic acid dissolved in aqueous alcohol—giving a dark green colour to fatty acid deposits.
(7) Mount in glycerin jelly.

Key questions
Why are only fatty acids stained? Are all the steps on this long-winded procedure really necessary? Does the copper have to be present as the acetate?

Mechanism
The presence of calcium ions in the fixative solution prevents losses of fatty acid anions by converting them into insoluble calcium salts ('calcium soaps'). Acidification of the section generates free carboxylic acids, which are virtually insoluble in water. In addition the acid extracts calcium phosphate deposited pathologically, which can otherwise give rise to artifactual staining. Treatment with the weakly alkaline copper(II) acetate solution generates water soluble fatty acid carboxylates, but immediately precipitates them as insoluble copper salts. To increase the colour intensity to a useful level, the copper(II) ions are then complexed with Rubeanic acid. To limit extraction of the non-ionic copper(II) Rubeanic acid chelate a hydrophilic mountant is used.

Small print. A wash with EDTA before Rubeanic acid complexation removes copper(II) ions loosely bound to other tissue sites, and so reduces background staining.

Implications
A negative staining outcome after the extraction control is essential for identifying fatty acid deposits—since any site binding copper(II), such as calcified tissues and phospholipid deposits, will stain with this method.

References
This method was originally described by Holczinger (1959). A critical study of the procedure, demonstrating the need for the acid pretreatment/acetone extraction control was carried out by Elleder and Lojda (1972).

Osmium tetroxide for unsaturated fatty acids

Objective

To stain unsaturated fatty acids, but not other lipids.

Staining reagent

Osmium tetroxide is a highly volatile, slightly hydrophobic compound; dissolved in water it gives an acidic solution. The oxidation state of the metal in osmium tetroxide is 8.

Critical technical step

(1) Immerse specimen in an aqueous osmium tetroxide solution—giving brown to black staining of unsaturated fatty acids; saturated lipids and cholesterol do not stain.

The reactivity of osmium tetroxide, both with lipids and with proteins, together with the compound's volatility, make it essential to work in a fume cupboard or similar device.

Key questions

Why are only unsaturated fatty acids stained? Why not cholesterol, since it too is unsaturated?

Mechanism

Initially the osmium(8) tetroxide reacts with carbon–carbon double bonds, forming colourless osmium(6) esters. Further reduction, and coordination of the reduced osmium with electron-donating ligands, can convert these esters to orange/brown/black 'osmium blacks'. Such reduction and coordination occurs for example with the thiol and amino groups present in proteins.

Small print. The failure of cholesterol to stain is not due to failure of the sterol's unsaturated carbon to form an osmium(6) ester, but to the subsequent non-reduction of the bound osmium. This perhaps reflects the relative chemical homogeneity of cholesterol deposits. Since osmium tetroxide does dissolve in saturated lipid droplets, artifactual staining can occur if this material is reduced by the action of tissue-processing fluids.

References

For a review of reactivity of osmium tetroxide with lipids see Hayat's monograph on fixation (1981). The role of protein in formation of osmium blacks has been investigated by Hanker and co-workers (Hanker *et al.*, 1967).

8.3.3 Stains for metal ions

In one of the two methods discussed, colour is generated by chelation of the metal by an organic ligand, and in the other by the precipitation of an inorganic salt.

Alizarin Red method for calcium

Objective

To detect low concentrations of calcium ions, usually in regions of pathological deposition.

Staining reagent

Alizarin Red S is a small, hydrophilic anionic dye, with a keto–hydroxy and hydroxyl–hydroxy grouping suitably oriented for chelating metal ions.

Critical technical steps

(1) Immerse section in slightly acidic, aqueous Alizarin Red solution for a short period, using microscopic control—giving orange-red staining of calcium deposits, with pale background.
(2) Blot, rinse in acetone, dehydrate through acetone–xylene.

Key questions

Why do calcium ions stain? Why does this acid dye, applied under acid conditions, give no substantial background staining?

Mechanism

Binding of Alizarin Red to calcium deposits is due to chelation of calcium ions by the dye, to give an insoluble coordination complex, with an overall charge of zero. Use of minimal staining times is necessary to prevent diffusion of calcium ions from the deposits; this tactic has the further advantage that background staining is reduced. Low background staining can also be attributed to the dye's small conjugated system and hydrophilic character not favouring non-coulombic contributions to affinity.

Small print. A study comparing the demonstration of calcium ions by a number of acid dyes indicated that high background staining correlated with large conjugated systems. Dissociation of the calcium chelate is minimized by avoiding aqueous post-staining washes.

Alternative views. Puchtler *et al.* (1969) considered that colouration of calcium is better understood as the formation of insoluble calcium salts of the Alizarin Red sulphonate.

References

The method has also been critically discussed by McGee-Russell (1958). Observations concerning influence of dye structure on chelation and background staining are found in Spiers (1970), though the interpretations are those of the present author.

Perl's Prussian Blue stain for ferric iron

Objective

To stain ferric iron, for example from haemosiderin pigment.

Staining reagent

The active component is the complex anion $Fe(CN)_6^{4-}$ present in potassium ferrocyanide.

Critical technical step

(1) Stain with an acidic aqueous solution of potassium ferrocyanide—giving blue staining of sites rich in ferric iron.

A positive control section should always be stained along with the test specimen, to establish the staining time needed. Analytical grade reagents should be used throughout the procedure, since most technical grade laboratory reagents are to some degree contaminated with iron.

Variants. Pre-staining treatments are necessary to demonstrate the ferric iron present in certain sites. High concentrations of acid are needed to demonstrate iron oxide, resulting from industrial exposure. If a concentrated hydrogen peroxide pre-treatment is used, then the iron present in haemoglobin and myoglobin may also be visualized.

Key questions

Why do only ferric ions give a blue colouration? Why do some sources of ferric ions require more vigorous pre-treatments than others?

Mechanism

Most ferric iron in normal tissues is complexed with proteins; in particular haemoglobin, myoglobin and the iron storage protein haemosiderin. As ferric iron is least tightly bound to the latter, the acid present in the ferrocyanide solution releases hydrated ferric cations from the haemosiderin. These cations then react with $Fe(CN)_6^{4-}$ anions to form an insoluble precipitate of Prussian Blue. More tightly bound ferric ions need more vigorous agents to release them as free reactive cations. Since ferrous iron does not form an insoluble ferrocyanide under acid conditions, it is not demonstrated.

Implications

Since ferric ions are highly soluble in water, procedures requiring pre-treatment with aqueous solutions of strong acid or hydrogen peroxide can result in substantial losses of iron from the tissues prior to the staining step.

References

This must be one of the earliest histo*chemical* procedures extant; Perls published his description of this method in 1867. For an account of the difficulties involved in its use, see the index entry 'iron' in the compendium of Lillie and Fulmer (1976).

8.3.4 A stain for polysaccharides

The periodic acid–Schiff procedure

Objectives

To demonstrate the presence of the 1,2-diol molecular fragment, and thus to visualize polysaccharides and other saccharide-containing biopolymers, such as GAGs and glycoproteins.

Staining reagents

The periodate anion is a water soluble oxidizing agent, with a high reactivity for 1,2-diol groups. Schiff reagent contains several essential components: a colourless, hydrophilic cationic derivative of Basic Fuchsin, with a sulphite anion covalently attached to the bridging carbon atom; plus hydrogen and sulphite ions.

Critical technical steps

(1) Oxidize with an aqueous solution of periodic acid.
(2) Immerse in aqueous Schiff reagent.

(3) Wash extensively in water—giving magenta staining of sites containing 1,2-diol groupings.

There are a number of *variants* for special applications. Thus to prevent extraction of those polysaccharides which are highly soluble in water, aqueous solutions can be avoided. The periodic acid is dissolved in alcohol; and the Schiff reagent can be precipitated as its phosphotungstate salt, which is applied to the specimen in alcoholic solution. Again, to permit staining of anionic polysaccharides the treatment with periodate is *either* grossly extended *or* carried out with a high concentration of magnesium chloride added. The Schiff reagent can be substituted by pseudo-Schiff reagents of various colours, prepared as is the type compound, namely by treating a cationic dye carrying a free amino group with acidic sulphite ions. Another substitute is to use Pararosanilin, dissolved in acid alcohol.

Key questions
What is the organic chemistry of the process? In particular, why is little colour generated until the final water wash? What is special about periodate, as an oxidant? Are all the features of the periodic acid–Schiff method explained by the organic chemistry?

Mechanism
The binding of the final coloured derivative involves a covalent bond between tissue and stain. However, Schiff reagent is a non-specific colour reagent for aldehydes, and the selectivity of the stain is due to the use of periodate. This oxidant reacts much more readily with 1,2-diols than with other molecular fragments, generating dialdehydes. Many other oxidants are less specific, attacking additional non-saccharide sites. Moreover they often convert aldehydes to carboxylic acids. The initial reaction product of tissue aldehydes with Schiff reagent is colourless. The colour results from the water washing removing the sulphite from the bridging carbon, and so generating a triphenylmethane chromophore.

Small print. Before any of these reactions can occur, the reagents must be present in the specimen. Hence affinity and rate factors relevant in dyeing are again of significance here. Thus anionic saccharides are only slowly oxidized by the periodate anion, though addition of salt allows a more substantial reaction. Moreover uptake of Schiff reagent, though colourless, is still controlled in part by its cationic character. Saccharide-free but anionic tissue sites, such as phospholipid-rich myelin sheaths and calcified sites, can take up Schiff reagent by ion-exchange. Then during washing a cationic dye is generated which, if not extracted, gives artifactual staining. Now consider the sensitivity of the method. In saccharide-rich sites such as glycogen granules, conversion of just a small percentage of the diol groups to aldehydes is sufficient to allow visualization.

Implications
Standardization of this method would require a consideration of all reagents and steps in the procedure: such factors as times and salt contents would have to be critically considered. Quantitation may be possible, but could be difficult in view of the fact that intense colouration can occur with only a small percentage reaction.

References
A number of general reviews of the periodic acid–Schiff procedure are available (for instance Cook, 1982). The influence of the electric charges of tissue components on periodate oxidation is discussed by Scott and Harbinson (1968). The dependence of sensitivity on degree of reaction was investigated by Ovadia and Stoward (1971). For an account of the staining mechanism of one of the Schiff reagent substitutes, namely Pararosanilin in acid alcohol, see Horobin and Kevill-Davies (1971).

8.3.5 Staining thiols

A coupled maleimide method

Objectives
To stain thiols, without staining disulphides or other molecular fragments.

Staining reagents
N-(4-aminophenyl)maleimide is a small hydrophobic compound which reacts vigorously with thiols to yield colourless thioethers. The aromatic primary amine on the maleimide is readily diazotized to form a reactive diazonium salt. Nitro Red is a small anionic dye, whose activating amino and hydroxy groups make it highly reactive with diazonium salts, yielding azo dyes.

Critical technical steps
(1) Fix tissue in buffered neutral formalin, containing the organomercurial Mersalyl; process through to paraffin.
(2) Treat dewaxed sections with acidified mercaptoethanol.
(3) React thiols with *N*-(4-aminophenyl)maleimide.
(4) Treat with nitrous acid.
(5) React with Nitro Red at pH 7.
(6) Differentiate in ethanolic triethylamine in 70% alcohol—giving selective blue colouration of thiol-containing sites.

This procedure is more sensitive than the standard method using dihydroxydinaphthyldisulphide, and more selective than methods such as the use of Mercury Orange.

Variant. The maleimide procedure can be extended to allow simultaneous demonstration of thiols and disulphides, as follows. Following step (6), the section is treated with tributylphosphine; steps (3) and (4) are then repeated, and the intermediate diazonium salt coupled with a nathol reagent such as chromotrophic acid, to yield an azo dye at the site of disulphides, of contrasting colour to that formed with Nitro Red.

Key questions
Why add Mersalyl to the fixative? What reactions occur on exposure of the section to mercaptoethanol, *N*-(4-aminophenyl)maleimide, nitrous acid, and Nitro Red? Why is the maleimide procedure both sensitive and selective? Since the coupling agent Nitro Red is an acid dye, what prevents background staining obscuring the colouration of thiol-rich sites? How are disulphides distinguished from thiols?

Mechanism
The highly reactive organomercurial, Mersalyl, converts free thiols to mercury derivatives. Consequently thiols are not lost by air oxidation to disulphides during tissue processing, nor by reaction with formaldehyde during fixation. Subsequent unmasking by mercaptoethanol releases free thiols. This preservation of native thiols increases sensitivity. The maleimide reacts with the thiols—but with nothing else—to yield colourless thioethers. Since the maleimide is both fast-diffusing and highly reactive, this reaction goes to completion—further increasing sensitivity. The thioether intermediates are visualized by first diazotizing them with nitrous acid, then azocoupling the resulting diazonium salts with Nitro Red, yielding a blue bisazo dye. The coupling rate is rapid, again aiding sensitivity.

Small print. Background acid dyeing is limited by the relatively high pH of the coupling process. Non-covalently bound dye is extracted using alcoholic triethylamine. The variant method, which allows the demonstration of disulphides, is based on the reduction of native disulphides to thiols by the strong reducing agent tributylphosphine, followed by repeating the thiol demonstration using a coupling agent giving rise to an azo dye of contrasting colour to that generated from Nitro Red.

Implications
Since this method is both sensitive and selective, it can provide the basis for a histochemical assay of thiols and/or disulphides.

References
This approach to thiol histochemistry is discussed in several papers by Sippel (for instance, 1978).

8.4 THE ENZYME HISTOCHEMICAL APPROACH

This strategy aims to generate immobile, visual labels at enzymic sites by using the catalytic properties of enzymes to modify applied substrates. To enable clear accounts of enzyme histochemical methods to be given, specialized information peculiar to these procedures is presented in section 8.4.1. Some of the general issues of this approach are then illustrated using vignettes of widely used methods.

8.4.1 The underlying processes of enzyme histochemistry

To achieve specificity the substrate should either be the native compound or a synthetic with very similar properties. The incubation medium should not diverge in pH and other significant properties from the native enzymic environment. Moreover the substrate must be present at the enzymic site at high concentration. Putting these needs together gives the following specification for the ideal substrate: it is fast diffusing, has high affinity for all tissue compartments, readily gives rise to a coloured or fluorescent product — and all this whilst still matching the chemistry of the native substrate!

In the real world contradictions between these requirements are numerous. For instance, some high-affinity substrates give substrate inhibition of the target enzyme; some specific substrates are poorly soluble in water, requiring addition of organic solvents which can reduce enzymic activity. Such problems are dealt with by compromise, on a case-by-case basis — there is no such thing as a free lunch.

To ensure effective staining, the final reaction products of enzyme histochemical methods are usually intensely coloured, or fluorescent; and must be closely localized around the enzymic site. These aims are sought in chemically quite different ways.

Some final reaction products are inorganic ionic crystals, such as lead sulphide; whose affinity for the tissue is due to coulombic stain–stain forces, not stain–tissue attractions. Other final reaction products are organic pigments, such as bis-formazans, which bind to tissue proteins by strong van der Waals forces. All reaction products must have a low affinity for the incubation medium, or other solvents to which the specimen is exposed after staining, and also a low affinity for tissue components distant from the enzymic site.

These latter requirements are again contradictory. Take for example reaction products which are organic pigments. If these have low affinities for tissue proteins, localization is poor; but high affinity can result in product inhibition of the target enzyme. Moreover, if they do not have low aqueous solubility, losses will occur into the incubation medium; but hydrophobic products may give artifactual staining of tissue lipids.

Such localized, coloured reaction products can arise directly from the action of the enzyme on a substrate. An important example is the generation of a coloured polymer from diaminobenzidine, or one of its derivatives, by the catalytic action of peroxidase. However, in many methods the enzymically catalysed reaction gives rise to an *intermediate* reaction product, which may be neither coloured nor immobile. Conversion to a suitable final product is then achieved non-enzymically using visualization and trapping reagents.

Visualization reactions are very varied, and include transformation of an immobile, but colourless, polymer into a coloured form, and conversion of a colourless lead phosphate precipitate into black lead sulphide. Trapping reactions include precipitation of mobile fatty acid carboxylate anions as their insoluble though colourless calcium salts, precipitating thiocholine as a colourless copper coordination compound, and many more. Some widely used methods achieve visualization and trapping simultaneously. In certain procedures for demonstrating hydrolases, a slightly soluble, colourless naphthol intermediate reaction product is converted to a brightly coloured, immobile azodye using diazonium salts as trapping/visualizing agents. In certain dehydrogenase demonstrations, an electron is removed from the cytochrome chain by a tetrazolium salt, generating a formazan pigment.

In these latter steps, technical contradictions can include the need for a visualization/trapping agent which is highly reactive, but not with the enzyme, and the need for a strongly adsorbed reaction product, but one which derives from a highly penetrating visualization/trapping agent.

Since successful use of the enzyme histochemical approach depends in large part on an appreciation of the necessary compromises, it is profitable to discuss two more chronic dilemmas. The first, arising prior to staining, is the conflict between localization and sensitivity. Localization, and indeed interpretation of cytological and histological morphology, requires active preservation measures, usually in the form of fixation. However, fixation denatures most proteins, and so inhibits many enzymes to a greater or lesser degree. Technical responses to this dilemma include both compromise (for example fixation at low temperatures, or for short periods) and use of non-fixative strategies (such as semi-permeable membranes), as mentioned in section 2.3.2. An analogous problem is how to maximize both sensitivity and specificity. If substrate is transformed too easily into product, this process may occur non-enzymically, reducing specificity. However, a weakly reactive substrate results in low sensitivity. These, and the earlier, problems emphasize the routine need for control methods in enzyme histochemistry. This technical requirement is too often – and too optimistically — overlooked. Controls must be varied, and take into account both enzymic and non-enzymic steps (see Chapter 5).

8.4.2 Examples from hydrolytic enzyme histochemistry

The examples illustrate the two most widespread strategies for demonstrating hydrolases, namely azodye and metal salt methods.

An azo-dye method for acid phosphatase

Objectives

To demonstrate the enzyme acid phosphatase, without staining the sites of other phosphatases.

Staining reagents

The substrate is a naphthyl phosphate: both α-naphthyl phosphate and various phosphates derived from the AS Naphthols are currently recommended in standard texts. The latter type carry amidophenyl substituents, and hence have larger conjugated systems than the simple naphthyl compound; they are moreover sometimes more hydrophobic. The visualizing agent is a diazonium salt. Three examples, from the wide variety currently recommended, are Fast Red TR, Fast Garnet GBC and Hexazonium Pararosanilin, which give rise to red azo dyes with most naphthols. All three diazonium salts are fast-reacting. The Pararosanilin derivative has the largest conjugated system. The first two can be stored as such, the cations being stabilized by large anions; the latter must be freshly prepared for each incubation from the primary amine.

Critical technical steps

(1) Fix tissue in formal–calcium, or in formaldehyde vapour; cut cryostat sections.
(2) Incubate at 37°C in a slightly acidic (pH 4.7–5.0), substantially aqueous solution containing the naphthyl phosphate substrate plus the diazonium salt — giving intense staining of sites rich in acid phosphatase.
(3) Mount in glycerin jelly, or, when Hexazonium Pararosanilin is used as visualizing/trapping agent, dehydrate quickly through alcohols to xylene, and mount in polystyrene.

The α-naphthyl phosphate is more soluble in water than most phosphates derived from AS Naphthols. Hence the latter compounds are typically taken up into a non-aqueous solvent such as dimethylformamide or dimethylsulphoxide. These solutions are then mixed with aqueous solutions of the other components of the incubation medium, the final solvent mixtures being more than 95% water. A wide variety of diazonium salts is available, allowing selection of a suitable colour for the final reaction product.

Key questions
What, precisely, are the intermediate and final reaction products? Why is the method selective for acid phosphatase? What are the dis/advantages of using substrates derived from AS Naphthols? Why use the, technically less convenient, hexazotized Pararosanilin?

Mechanism
The enzymically catalysed step is the hydrolysis of a naphthyl phosphate ester, yielding as initial reaction products a naphthol plus inorganic phosphate. Acid phosphatase is not a very substrate-specific enzyme, hence its acceptance of the synthetic naphthyls. However, it is pH-dependent, allowing its selective demonstration by use of a low pH incubation medium. AS Naphthols, owing to their relatively large conjugated systems and hence stronger van der Waals attractions, have a greater affinity for tissue proteins than does α-naphthol. Since the simple naphthol is in fact slightly hydrophobic, its use can result in artifactual staining of tissue lipids. Conversion of naphthols to azo dyes by reaction with a diazonium salt is thus merely a visualization procedure with the AS Naphthols, but is both visualization and trapping with α-naphthol. To limit diffusion of the intermediate reaction product a rapid azocoupling reaction is desirable; and indeed all the three diazonium salts named above are highly reactive. To minimize diffusion of the final, coloured reaction product a large conjugated system is an advantage. Hence the use of Hexazonium Pararosanilin. The strongly adsorbed reaction products resulting from this reagent are also relatively hydrophilic. The overall effect gives good localization and permits processing stained specimens through alcohols and xylene into a polystyrene mountant. Use of the diazonium salts with smaller conjugated systems yields hydrophobic azo dyes with lower affinities for tissue proteins. Sections thus stained must be mounted in aqueous media to avoid extraction of the dye.

Small print. α-Naphthyl phosphate does have advantages over the AS Naphthol products: it is cheaper, and usually much purer.

Implications
By choosing substrates and trapping/visualizing agents with suitable structures, final reaction products with the desired colours and localizations can usually be achieved.

Metal ion precipitation method for adenosine triphosphatase

Objective
To demonstrate adenosine triphosphatase (ATPase) whilst not staining the sites of other phosphatases. ATPase staining is used routinely in histopathological investigation of skeletal muscle biopsies.

Staining reagents
The trapping and visualizing agents are inorganic salts, whose active constituents are calcium, cobalt and sulphide ions.

Critical technical steps
(1) Cut unfixed cryostat sections.
(2) Optional: expose sections to pH 9.4, 4.6 and 4.2 solutions.

(3) Incubate in a mildly alkaline aqueous solution containing the substrate adenosine triphosphate, plus calcium ions.
(4) Treat with an aqueous solution of cobalt chloride.
(5) Treat briefly with an aqueous solution of ammonium sulphide — giving black staining at sites of ATPase activity.
(6) Dehydrate through the alcohols to xylene, and mount in polystyrene.

This complicated method requires careful removal of reagents after each step. However, such washing can be overdone. Even the supposedly insoluble lead sulphide final reaction product is in fact slightly soluble, especially when washing is carried out in weakly acidic distilled water. The calcium ion concentration must not be too low, or artifactual staining of non-enzymic sites occurs. The optional pre-incubation treatments with controlled pH solutions are useful when performing muscle biopsies; since the different pHs activate/deactivate the ATPase variants present in the different fibre types of human skeletal muscle.

Key questions
What is the enzyme-catalyzed process? Exactly what are the intermediate and final reaction products, and what are the functions of the trapping and visualization reagents? What are typical artifacts with this procedure?

Mechanism
The ATPases are sensitive to fixative agents and so, since they are not readily lost from the cell cytoplasms, staining is usually carried out on unfixed material. The enzymic step is the hydrolysis of the specific substrate, adenosine triphosphate, to yield the intermediate products adenosine diphosphate plus inorganic phosphate. This latter ion is trapped by being precipitated as calcium phosphate. For this reason, low calcium or phosphate ion concentrations, however they arise, result in the phosphate not immediately precipitating and so not being confined to the enzymic sites. Artifacts resulting include diffuse staining, and staining of cell nuclei. Precipitated calcium phosphate is colourless, and visualization is achieved via two ion-exchange steps. In the first, calcium cations are exchanged with cobalt cations; in the second, phosphate anions are exchanged with sulphide. The resulting black cobalt sulphide is clearly visible.

Implications
The concentration of enzyme varies from one site of occurrence to another; moreover the permeability of the sites vary. Consequently the rates at which phosphate ions are released, and calcium ions arrive, vary too. Consequently the sensitivity and localization achieved with such methods are site-dependent.

Phosphorylase demonstration by amylose or glycogen formation

Objective
To demonstrate amylophosphorylases activity; a common histopathological application being in fibre-typing muscle biopsies.

Staining reagents
The visualizing agent is an iodine/iodide solution, in which the active constituent is the triiodide anion (see section 8.1.7).

Critical technical steps
(1) Cut unfixed cryostat sections.
(2) Incubate in a slightly acidic aqueous solution containing glucose-1-phosphate, adenosine triphosphate, glycogen, and magnesium chloride; an optional addition is an enzyme inhibitor – ethanol, or fluoride or mercuric ions.
(3) Rinse in aqueous ethanol; then fix briefly in absolute ethanol.
(4) Visualize the reaction product using a iodine/iodide solution — giving staining of those sites of phosphorylase activity: blue/black if an inhibitor was present in step (2); brown/purple if not.
(5) Mount in a mixture of iodine/iodide solution and glycerine jelly.

Some *variants* add the colloid stabilizer polyvinylpyrrolidone to the incubation medium. If permanent preparations are wanted, alternative mountants are called for: for example iodine/iodide plus Apathy's gum syrup.

Key questions
What is the enzyme catalysed reaction? How is the reaction product visualized?

Mechanism
Phosphorylase is fixative-sensitive, so its demonstration is carried out on unfixed specimens; loss of enzyme from such material can be inhibited by adding a colloid stabilizer (see section 2.3.2). *In vivo* the phosphorylase is usually engaged in catalysing the hydrolysis of amylose or glycogen to generate glucose. However, the enzyme will also catalyse the reverse, synthetic, reaction and it is this which is exploited histochemically. Thus the enzyme catalysed step in this procedure is the polymerization of glucose-1-phosphate. The initial product is a straight chain amylose, but when the inhibitors are absent this is acted on by a branching enzyme, yielding glycogen. These phosphorylative polymerizations are magnesium-activated, and require adenosine triphosphate as a phosphate source. Some pre-formed polymer is also required as a primer. There is sometimes sufficient glycogen present in the tissues for this; but since this is not always so glycogen is routinely added to the incubation medium. The reaction products amylose and glycogen are both extremely water-soluble polymers, but are not soluble in alcohol. Hence the partially ethanolic solvent used for the post-incubation washing, antd the ethanolic post incubation fixation. Visualization of amylose or glycogen is achieved by the formation of triiodide complexes, which gives blue and red/brown products respectively (see section 8.1.7).

Implication
As glycogen is a normal constituent of those structures containing the phosphorylase, the staining of an unincubated control section with the iodine/iodide reagent is essential.

Tetrazolium salt method for succinate dehydrogenase

Objective

To demonstrate sites of succinate dehydrogenase (SDHase).

Staining reagent

The tetrazolium salts most commonly used as visualizing/trapping agents are Thiazolyl tetrazolium (MTT), Nitro Blue tetrazolium (NBT) and Tetranitro Blue tettrazolium (TNBT). The former is a monotetrazolium salt, whilst the latter two are bis-tetrazolium salts whose conjugated systems are more than twice as large. Reduction of these salts give hydrophobic formazans. The MTT formazan gives a 2:1 MTT–cobalt coordination complex with cobalt(II) ions.

Critical technical steps

(1) Cut cryostat sections: either use these directly, or after brief fixation.
(2) Incubate sections in an aqueous medium containing succinate and a tetrazolium salt, plus Co(II) ions when using MTT — giving localized staining of the sites of SDHase activity: black with MTT-cobalt, blue with NBT, and red-brown with TNBT.
(3) If section not fixed in step (1), post-fix.
(4) Mount sections stained with MTT–cobalt in glycerine jelly; sections stained with NBT or TNBT can be taken through the alcohols and xylene, and mounted in polystyrene.

Several fixative regimes have been used successfully: for instance brief fixation with formaldehyde, use of acetone, use of low temperatures. Addition of the colloidal stabilizer polyvinylpyrrolidone (PVP) to the incubation medium has also been used (see section 2.3.2).

Key questions

What is the enzymic step? How does trapping/visualization occur? How do the tetrazolium salts differ?

Mechanism

SDHase is fixative-sensitive. Pre-staining fixation is either avoided, and enzyme loss reduced by addition of PVP, or minimized in terms of reagent and conditions. Alternatively post-staining fixation is adopted to preserve morphology, and loss and diffusion of enzyme is accepted. The enzyme-catalysed step is the dehydrogenation of succinate, with transfer of hydrogen ions to the enzyme itself, and electrons to the cytochrome carrier system. Regarding protons and electrons as the intermediate reaction products, both trapping and visualization is achieved with a tetrazolium salt. These colourless ionic compounds, dissolved in the incubation medium, are reduced to coloured, non-ionic formazans which precipitate around the enzymic sites. NBT and TNBT yield bis-formazans, whose large conjugated systems favour stain–tissue protein van der Waals attractions. Consequently these formazans give finely granular, or even non-crystalline, staining. Even though the formazans are hydrophobic, lipids are not stained, as there is insufficient non-adsorped formazan. MTT yields a hydrophobic monoformazan, with a small conjugated system. This is not strongly adsorbed and can stain lipids and form large, poorly localized, crystals. However, addition of cobalt(II) ions to the incubation medium results in the precipitation of a MTT formazan–cobalt complex. This complex is more hydrophilic and has a large conjugated system. However, the complex is still smaller than NBT and TNBT, so that only the latter have sufficient affinity to resist extraction by alcohol and xylene. Such structural influences of the various tetrazolium salts have been discussed elsewhere (Horobin, 1982).

8.5 IMMUNOSTAINING, *IN SITU* HYBRIDIZATION, AND OTHER AFFINITY HISTOCHEMICAL STRATEGIES

These approaches involve probes which are naturally occurring molecules — or synthetic mimics of such natural products — which possess selective affinity for cellular targets. To such probes are added, directly or indirectly, visual labels. The combination of probe-plus-label then constitutes the essence of the staining system. Unlike most stains discussed previously, the reagents involved are often macromolecular. However, the essential characteristic of this group of methods is not so much their chemical nature as the fact that their selectivity was not designed but has *evolved.* Moreover in many cases the staining reagents themselves are not synthesized in a laboratory but in an organism; although the visualizing systems are added synthetically.

The layout of this section is familiar. A sketch of the peculiarities of affinity histochemical methods, in section 8.5.1, is followed by several vignettes, describing varied methods which illustrate the underlying principles.

This introductory account is biased towards labelled antibodies, even though many of the concepts apply to other reagent systems. This bias reflects the current practical predominance of immunological methods, with the consequent study of their mechanisms and problems. Moreover, even methods whose fundamental selectivity involves an interaction other than that of antibody with antigen may be visualized with the aid of labelled antibodies!

8.5.1 Underlying processes of affinity histochemistry

A wide variety of target–probe interactions have been used histochemically, as illustrated in Table 8.1. Note that, except with nucleic acids, either the target or the probe — or both — are proteins.

The overriding advantage of using these evolved reagents is the enormous range of highly selective probes which exist. For instance, a mammal such as *Homo sapiens* makes perhaps one hundred thousand different antibodies. So if we wish to detect a micro-organism infecting human tissues, there is a good chance that an immunoglobulin synthesized by a rabbit or rat will bind with some epitope on the bacterium. Sometimes indeed one can imagine no other way a particular entity could be detected except by an evolved interaction. For instance, how else would a receptor for a particular hormone, present in the plasmalemma of a cell, be detected except by exploiting the hormone–receptor affinity?

What kind of physicochemical factors underlie such highly selective interactions? And why are so many different selectivities available? The influences behind these selective interactions are very familiar: electric charge, van der Waals attractions,

Table 8.1 — Examples of 'evolved selectivities' exploited to provide staining selectivity in affinity histochemical procedures

Biological target	Probes used
Antibody	Antigen
Antigen	Antibody
Biotinylated biopolymers	Avidin
DNA sequence	Complementary DNA or RNA
Enzyme	Specific inhibitor
Enzyme substrate	Enzyme
Fc portion of immunoglobulins	Complement, Protein A
Hormone receptor	Hormone
Motor end plate receptors	Specific toxin

hydrophobic effects, hydrogen bonding. Selectivity arises from the complex three-dimensional patterns of such influences arising within macromolecules. The simplest are perhaps the binding of complementary strands of nucleic acids. Here the binding of an RNA or DNA probe to a sequence in the chromosomal DNA is largely due to the fact that certain base-pairs form strong hydrogen bonds. Hence high affinity depends on a very specific sequence of bases. When one or more of the interaction partners is a protein, an interacting surface will consist of a three-dimensional pattern of varying charge, polarizability, hydration, hydrogen bonding potential and so on. The chances of such a surface matching a cellular target by chance alone is trivially small. The existance of large numbers of such interactive systems arises from the fact that we are here concerned with the chemical language of the cell, which requires many selective interactions of many kinds.

To exploit such interactive precision in a histochemical context, it is necessary to couple a probe to a visualizing system. The possible nature of the system, and indeed of the coupling, are varied.

Conceptually the simplest is when a label is directly attached to the probe. Thus a fluorochrome, such as Fluorescein isothiocyanate, may be attached covalently to an RNA strand, to detect complementary DNA sequences, or attached to an immunoglobulin to detect an antigen, or attached to an enzyme to detect a substrate. The same strategy can be used in electron microscopy, but now the label must be electron scattering. Two familiar examples are the labelling of probes with colloidal gold or with the iron metalloprotein ferritin. Finally note that it is possible to attach objects of characteristic shape to a probe as labels. Examples of such labels are erythrocytes, for use with the light microscope, and cylindrical virus particles with electron microscopy.

However, the visualizing system may be indirect, either in its mode of action or in its method of attachment to the probe. Thus the probe may be directly labelled by an invisible agent, which agent, however, may then be visualized, or used to generate a visible substance. An example is enzyme labelling, most often with peroxidase. The presence of this label can be demonstrated by incubating the specimen (containing the target–probe–peroxidase complex) in diaminobenzidine (DAB) plus peroxide. The enzyme catalyses the formation of a brown DAB-polymer which, since it is reactive with osmium tetroxide, can also be used in electron microscopic investigations. An interesting varient is the use of silver impregnation to intensify staining due to a colloidal gold-labelled probe. Though even single particles of gold may be seen in the electron microscope, direct gold staining can only be used with light microscopy when large amounts of gold are bound. However, since colloidal gold catalyses the reduction of silver(I) to metallic silver, optically visible silver staining can be generated from weak gold labelling (see section 8.2.3). Note that such silver intensification can also be applied to peroxidase-labelled probes, since the DAB-polymer also possesses catalytic properties.

A visualizing system may also be coupled indirectly to a probe. Consider an unlabelled probe binding to a target in a specimen. If a second probe, with selective affinity for the first, is now labelled it may be used as an indirect stain for the target. Inspection of Table 8.1 indicates several variants of this strategy. One in particular is currently in widespread use: Sternberger's PAP immunostaining procedure. This involves the sequence of interactions shown below, all of the evolved type. Note: Ig is shorthand for immunoglobulin:

tissue	rabbit	swine	peroxidase-rabbit
antigen(X)——	(anti-X)Ig——	(anti-rabbit Ig)Ig——	(anti-peroxidase)Ig
			complex

Here the label is the fourth protein in a row, from the target.

Compared to other forms of staining reagents, methods for obtaining some of the evolved reagents used in affinity histochemistry are unusual, namely, persuading live cells or organisms to synthesize the reagents for our use. Once the probes are made, and purified — which latter process is often difficult — procedures such as attaching labels are more orthodox, although not without their own problems.

Admittedly the evolved reagents are sometimes obtained by straightforward, if painstaking, biochemical processing of dead tissues. Examples of this include the Complement proteins, and enzymes. Indeed some reagents of this class can be prepared synthetically, e.g. steroid hormones and some polynucleotides. However, with immunoglobulins a live organism, or cell culture, must be stimulated to produce the particular reagent we require. For this a sample of the appropriate target, of sufficient purity for the problem in hand is needed. Moreover the sample must be in a suitable physicochemical form to be antigenic. Meeting these specifications may call for purification — which can prove difficult when the antigen is poorly characterized biochemically — and also the adsorption of the target material onto a particulate or polymeric material in order to enhance antigenicity. Challenge of the immune system of a suitable animal with such an antigen then results in plasma cells synthesizing the appropriate immunoglobulin. The protein may be separated from serum samples, and purified by such procedures as immunoprecipitation and affinity chromatography.

Once such reagents have been obtained their labelling may be achieved in various ways. In the peroxidase–antiperoxidase complex, widely used in immunostaining, the label–reagent binding is itself immunological. Contrasted to this is the fluorescent, and some enzyme, labelling of immunoglobulins and other reagents. In these methods the labels are attached covalently by reaction of isothiocyano or chlorosulphono groups on the fluorochromes with nucleophils on the proteins. Such processes must be carefully controlled. Firstly they must not cause denaturation of proteins, or no useful probe will exist. Moreover the ratio of label to evolved reagent is important. A low label–reagent ratio often results in poor sensitivity. However, since attaching label can alter the binding properties of proteins, overlabelling must also be avoided. Whatever the means of labelling, the final product must contain neither *un*labelled reagent nor unattached label. Free label may bind to the specimen to give false positive staining; and unlabelled reagent will compete with binding of the labelled form, so reducing staining intensity.

Achieving specificity, or at least high selectivity, with affinity histochemistry typically involves two different aspects. First achieving a selective interaction between the evolved reagent itself and the target; and second ensuring that non-biological factors neither prevent nor simulate reagent–target interactions.

Probe-target selectivity of evolved reagents does in fact vary considerably. Thus some steroid receptors will bind a wide range of small hydrophobic molecules, whilst enzymes may be specific for a single substrate, or merely weakly selective. Antibody–antigen interactions are generally regarded as specific; but even here qualifications must be made. Obtaining pure antisera may be difficult — contamination by other antigenic material, e.g. intruding micro-organisms, can occur. Second, a pure antiserum will be polyclonal, that is it will contain a range of immunolglobulins, targeted against different molecular aspects of the antigen. Of course it is now possible to prepare monoclonal antibodies, containing only single immunoglobulin species, which have as target single molecular fragments of the antigen. Even with such monoclonal antibodies a type of non-specifically can arise when dealing with protein antigens: the same molecular fragment may be found on more than one protein. So a reagent regarded as 'anti (protein X) immunoglobulin' may in fact also bind to protein Y and protein Z, when the same epitope occurs on all three species. As a tailpiece, note that antibodies may sometimes bind non-immunologically to tissue proteins, e.g. to proteins of myelin sheaths; and analogous effects may arise with other evolved reagents.

One common non-biological cause of staining failures is the inaccessibility of cellular targets to the reagents. Protein reagents are, in molecular terms, large — and can be enormous. Think of a peroxidase–antiperoxidase complex which typically contains two immunoglobulin and three peroxidase molecules, whose overall molecular mass is hundreds of thousands of daltons. Access of such species is limited to the free surfaces of specimens. In cultured cells surface receptors can be detected directly, whilst demonstration of internally sited antigens requires a preliminary permeabilization with surfactant or protease. Analogously, cryostat sections — which have been through the shattering freeze–thaw cycle — are permeable, whilst staining of plastic embedded material is limited to the surface layers.

Postive staining artifacts due to non-biological binding also occur. Some effects of this type arise from the nature of the label. Certain enzyme labels, for reasons which are not entirely clear, can bind to various cell and tissue sites. Fluorescent labels can also enhance non-immunological staining of labelled antibodies. This is probably due to the large conjugated systems of the fluorochromes giving rise to strong van der Waals attractions. However, coulombic effects have also been demonstrated in immunostaining. The isothiocyano and chlorosulphonyl groups of the common fluorescent labels react with the amino groups of antibodies, leading to the proteins becoming more anionic. Consequently the labelled reagents act like giant acid dyes. In keeping with this, non-immunological staining is most pronounced in regions rich in basic proteins, and is suppressed by formalin fixation or use of alkaline staining baths. When evolved reagents are enzyme-labelled, or when silver intensification of staining is used, then artifacts typical of enzyme histochemistry and silver staining can arise.

When using routine methods, achieving sensitivity is almost as important as achieving selectivity. Obvious factors influencing sensitivity are accessibility and affinity of target to probe. These factors can sometimes be favourably manipulated, e.g. immunostaining of formalin-fixed specimens is often enhanced by their prior treatment with proteases. When achieving sufficient sensitivity is a limiting factor, monoclonal antibodies have an intriguing *dis*advantage. A monoclonal antibody will bind to a single antigenic determinant on an antigen; whereas polyclonal antibody preparations contain a number of immunoglobulins, each of which binds to a different determinant. So increased selectivity is achieved at the price of a reduction in sensitivity. Enhancement of sensitivity in affinity histochemistry is often achieved by the use of indirect staining methods. Indirect labelling means that the primary probe is unlabelled, but itself becomes the target for several labelled secondary probes, giving a marked gearing up effect. Indeed, as in the peroxidase–antiperoxidase immunostaining technique, more than one amplification step can be used. Note that such methods enhance the number of labels bound per target. However, another strategy is to enhance the amount of, or visibility of, stain produced at each target site. An obvious example arises with enzyme labelling. Extension of the enzymic incubation time increases the amount of reaction product formed, and so the sensitivity. Alternatively an enzyme substrate may be chosen such that the reaction product is fluorescent. Yet another useful variant on this theme is to use the DAB-polymer induced by peroxidase labels to catalyse a silver staining reaction.

The other pay-offs from indirect staining methods are manipulative convenience and financial savings. Compare direct and indirect immunostaining methods. With direct methods a labelled antibody has to be produced for each antigen of interest. This means that, except for antigens of general diagnostic concern, costs will probably preclude availability of commercial materials. However, with indirect methods only one labelled antibody is required, which ensures commercial viability.

Finally there is — as in the discussion of enzyme histochemistry — a plea for the routine use of control procedures. Whenever possible these should include: omission of the probe, or of the primary probe for an indirect method; pre-absorption of the probe solution with a pure sample of the target; replacement of the probe with some generally similar compound that, however, has no specific affinity for the target; and staining of a known positive specimen. Methods using enzyme labelling or silver intensification should also add controls relevant to these visualizing steps.

8.5.2 Illustrations from routine methods

Direct immunofluorescent staining of antigens

Objective

To demonstrate various antigens. Some current diagnostic applications include: detection of IgA-producing plasma cells in the bone marrow of patients with an IgA myeloma; detection of antibody deposition in the glomerulus of a patient with membranous glomerulonephritis; and the detection of the C3 Complement component at the dermal/epidermal junction in a patient with pemphigoid.

Staining reagent

The relevant antibody is obtained, and labelled with a reactive fluorochrome such as FITC. This is a derivative of Fluorescein, carrying an isothiocyano group which reacts with various nucleophilic groups on proteins. Amino groups on immunoglobulins will react for instance, binding a fluorescenyl residue via a thiourea bridge. Such a derivative is anionic, slightly hydrophilic, and has a conjugated system of moderate size. The dianionic form of the derivative is particularly fluorescent.

Critical technical steps

(1) Cut cryostat sections of unfixed material.
(2) Wash well in phosphate buffered saline.
(3) Incubate with the FITC-labelled antibody, at the greatest dilution which still gives strong staining of the antigen.
(4) Wash well with phosphate buffered saline—giving selective fluorescence of antigenic sites.
(5) Mount in a slightly alkaline aqueous solution.

On unfixed cryostat sections, immunofluorescence is considered to yield better resolution than will peroxidase labelled antibodies. If the final preparations are cover-slipped, and these sealed in place with nail varnish or other medium, the preparations may be stored in the refrigerator for weeks or months. Longer storage times may be accommodated using polyvinylalcohol as mountant, with storage in the deep freeze.

Key questions

How is antibody–antigen binding maximized, whilst minimizing non-immunological staining? Why use an alkaline mountant?

Mechanism

When staining with a labelled anti(X) antibody, we may expect the labelled immunoglobulin to bind with high affinity to any of the antigen X which is present. The labelled antibody may also bind non-immunologically to various sites. This will be due in part to the presence of the label. The conjugated system of the Fluoresceinyl label will favour reagent–tissue van der Waals attractions; and the fluorochrome's anionic character will give rise to acid dyeing of tissue proteins. In order to minimize such binding, the Fluoresceinyl–antibody molar ratio is kept low, by removing heavily labelled antibodies chromatographically before staining. For the same reason the labelled antibody is used at the highest dilution still permitting staining of antigen; and the stained section is washed extensively with saline. The efficacy of both these procedures rests on the fact that the antibody–antigen binding is of higher affinity than is the non-immunological staining. The saline wash prior to exposure of the section to labelled antibody is considered to remove proteins which might otherwise block antibody–antigen binding.

Small print. The alkaline mountant is used to maximize the fluorescent intensity of the label.

Implications.
The procedures used in direct immunofluorescent methods such as this have changed very little since their introduction more than thirty years ago. It is therefore instructive to compare such methods with more modern indirect procedures, such as the one using the peroxidase–antiperoxidase complex described below. Direct immunofluorescent procedures are technically simple, and numerous labelled antibodies are commercially available. However, they are insensitive, and so are unsuitable for detecting low concentrations of antigen.

Indirect staining of antigen using PAP complex

Objectives
To demonstrate antigens of interest, especially when the specimen has been previously fixed and embedded. The range of specific applications is wide and widening. However, major current histopathological applications are: the demonstration of tumour markers, including hormones, immunoglobulins, and other classes of compound; identification and localization of micro-organisms; and localization of enzymes.

Staining reagents
The labelled entity is a peroxidase–anti(peroxidase) immunoglobulin complex (that is, a PAP complex). This is visualized by an enzyme histochemical demonstration of the peroxidase, in which process the visualizing agent is most commonly diaminobenzidine (DAB).

Critical technical steps
Assume, for clarity, we are here demonstrating the hormone gastrin. Note the use of Ig as the standard abbreviation for immunoglobulin.

(1) Treat dewaxed sections, cut from formalin fixed specimens, with methanolic hydrogen peroxide.
(2) Hydrate the sections; and optionally treat with a proteolytic enzyme (e.g. trypsin).
(3) Incubate the sections in normal swine serum.
(4) Incubate sections with the primary antiserum, namely rabbit anti(gastrin)Ig. Serial sections are treated with a range of immunoglobulin concentrations; the solutions all contain an excess of normal swine serum, as do all the other antisera.
(5) Incubate sections in swine anti(rabbit Ig)Ig.

(6) Incubate the sections in PAP complex.
(7) Incubate in DAB medium—giving brown staining of antigenic sites.
(8) Optional silver intensification procedure—converting even optically unnoticed yellow specks into black staining.

The optional treatment with a proteolytic enzyme, in Step (2), permits the wider use of formaldehyde fixation, since after protease treatment antigen is demonstrated which is otherwise undetectable. However, if the enzymic treatment is over-extended, non-specific background staining is also enhanced. Optimum enzyme extraction times can vary from tissue to tissue. Non-immunological staining is also reduced by the use of non-specific swine serum, both its addition to all antisera (Steps (4) and (5)), and its use as a pre-treatment agent in Step (3). The primary antiserum is applied to serial section, using a range of concentrations. Whenever possible, very dilute solutions are used, since this further reduces background staining, and also prevents the steric inhibition phenomenon giving false negatives. One important procedure not mentioned above is the careful washing of the sections after each incubation step. If such washings are not effective, antigen–antibody complexes may be precipitated onto the sections, complicating interpretation of the final image by generating false positives. The use of DAB to visualize the peroxidase enzyme label is a possible health hazard. There are alternative visualization systems, e.g. the α-naphthol/basic dye method, or the use of aminoethylcarbazole. The major *advantage* of the indirect PAP complex method is its high sensitivity. This is why it can be used after formaldehyde fixation. Even if only a few antigenic sites survive, visible staining can still be obtained, especially if the silver intensification procedure is used. This high sensitivity has the further advantage of allowing the use of extremely dilute primary antisera, which both reduces background staining and saves money.

Key questions

What is the immunological basis of this complicated procedure? What is the—seemingly illogical—basis of reducing false negative staining by *lowering* the concentration of the primary antiserum? Why does non-specific swine serum, applied with specific antisera and as a pre-treatment, also reduce background staining? What is the basis of the DAB visualization procedure for the peroxidase enzyme label? Why pretreat with methanolic peroxide? How does the silver intensification, in Step (8), work? What is the advantage of the PAP complex, compared to the various peroxidase–immunoglobulin conjugates use previously?

Mechanism

The various immunoligical interactions are as follows. The target antigen, e.g. the hormone gastrin, which survives fixation will form antigen–antibody complexes with the rabbit anti(gastrin)Ig present in the primary antiserum. This rabbit immunoglobulin then in turn forms an antigen–antibody complex with the swine anti(rabbit Ig)Ig of the secondary antiserum. Finally this swine antibody also binds to the rabbit anti(peroxidase)Ig of the PAP complex. If, after attempting to visualize the peroxidase, no staining is seen at sites expected to be gastrin-rich, several questions arise. First was antigen inactivated during fixation? If so, use of an unmasking protease extraction in Step (2) may facilitate staining. Instead of or as well as this, a milder fixative regime may be explored, e.g. a lessened exposure to formaldehyde, or use of vapour phase fixation with an agent such as benzoquinone. Another possibility is the occurrence of false negative staining, owing to the primary antiserum being too concentrated. This phenomenon involves the deposition of a very high concentration of rabbit immunoglobulin at the antigenic site. This can result in all the Fab sites of the secondary swine antibody being bound to the primary rabbit immunoglobulin, so that no subsequent binding occurs of the rabbit immunoglobulin present in the PAP complex. This possibility can be checked by using a more dilute primary antiserum. Positive artifacts can also occur, for instance immunoglobulins bind non-specifically to tissue sites such as connective tissue fibres. This binding can be masked by ensuring that an excess of non-specific immunoglobulin is present through the staining procedure. In the present method non-specific swine serum is used in this way; as an economical source of non-specific immunoglobulin. The peroxidase label is visualized using an enzyme histochemical procedure. This involves incubating the section together with DAB hydrogen peroxide. Enzymically generated peroxide radicals react with the DAB yielding a brown, polymeric reaction product, which is often directly visible. However, if the stained preparation is treated with a physical developer, the polymer can catalyse the reduction of silver(I) to metallic silver(0). This is the chemical reaction underlying the silver intensification procedure of Step(8) (see section 8.2).

Small print. The initial treatment of sections with methanolic peroxide serves to inactivate endogenous peroxidases and other compounds with peroxidase activity, eliminating one possible positive staining artifact. Note that the immunological attachment of peroxidase to antibody in the PAP complex minimizes non-immunological background staining, whereas the relatively hydrophobic bridging agents of previous reagents enhanced such artifactual staining.

Visualizing DNA sequences using* in situ *hybridization with labelled RNA

Objective

To identify specific DNA sequences, e.g. Adenovirus-5, or the mitochondrial DNA of *Crithidiae luciliae.*

Staining reagents

Complementary RNA sequences were prepared by standard molecular biological methods. These RNAs were oxidized with periodate, to yield dialdehydes at the 3′-terminals. Fluorescein isothiocyanate, or Tetramethylrhodamine isothiocyanate, were converted to their thiosemicarbazides and bound to the aldehydes to give fluorescent labelled RNAs.

Critical technical steps

(1) Treat target cells with acetic acid–methanol, then with dilute aqueous sodium hydroxide.
(2) Stain the cells with the labelled RNA in aqueous formamide solution, for very extended periods in the dark at room temperature.
(3) Rinse several times for extended periods in aqueous formamide in the dark—giving selective fluorescence of the structures containing the target DNA sequences.

Key questions

Why treat with acetic acid-methanol, or use the alkaline post-treatment? Why is it necessary to stain for very long times at room temperature in the dark? What is the need for the extended rinsing periods?

Mechanism

The initial treatment with acetic acid–methanol fixes the DNA by insolubilizing associated proteins. The slightly acidic solvent prevents dissolution of the DNA before the protein denaturation is completed, without hydrolysing the nucleic acid. The subsequent alkali treatment denatures the DNA, allowing access of the fluorescent probe to the resulting single-stranded polynucleotide chains. Exposure of the cells to the probe give rise to extensive fluorescent staining—owing to hydrization of the labelled RNA with complementary DNA sequences, and also to non-specific uptake onto various other entities. Since hybridization is a sterically controlled process involving a staining reagent which is macromolecular, this process is slow. Moreover it cannot be thermally accelerated without unacceptable side reactions. Once uptake of probe is complete, differentiation removes the less tightly bound non-hybridized stain. Again diffusion of the macromolecular RNA is slow.

Small print. The slow staining and de-staining steps are carried out in the dark to avoid photoactivated side reactions induced by the fluorescent labels.

References.

Synthesis and use of fluorescent-labelled RNAs was described in two papers from van Duijns laboratory (Bauman *et al.*, 1981a and b). Many other probes, not all RNA-based, have been developed; several reviews already exist (e.g. Bauman *et al.*, 1984).

Demonstrating β-adrenergic receptors using a fluorescent-labelled ligand

Objective

In the specific application described here, the object is to identify any β-adrenergic receptors present on bronchiolar epithelial cells.

Staining reagent

The probe used to detect the receptors is Propranolol; which is labelled by attachment of a fluorescent 9-aminoacridyl moiety.

Critical technical steps

(1) The fluorescent probe is administered intravenously to a live animal.
(2) The animal is killed and the airway dissected out. Portions are sectioned on a cryostat, and the sections mounted and air-dried.
(3) These sections are observed in a fluorescent microscope—discrete, bead-like areas of fluorescence occur on otherwise non-fluorescent Clara cells.

As a *control* unlabelled Propranolol is administered to the animal prior to the labelled form—if staining is specific, it will be inhibited by this blockade.

Key questions

Why can the drug Propranolol be used as a specific β-adrenergic receptor probe? Why does attachment of the basic fluorochrome not result in artifactual staining of nucleic acids?

Mechanism

Propranolol was a drug developed by pharmacologists to interact selectively with β-adrenergic receptors. This selectivity is retained in the fluorescent-labelled derivative, hence the efficacy of the staining procedure. Since the probe is administered to a live animal the surface receptors are accessible, but cationic label cannot reach the intracellular DNA and RNA.

Reference

This particular receptor-detecting method was developed by Masaro and Davis (1984).

9

The selection and design of staining methods

9.1 SPECIFYING THE PROBLEM

If you wish to know 'Are there any IgA secreting plasma cells in the bone marrow?' or 'Is the distribution of elastic fibres around this blood vessel normal, or does it show signs of pathology?' or 'Is the proportion of helper T-lymphocytes in this spleen within the normal range?' you are dealing with questions which concern pathologists, who are major users of histochemical methods. Consequently procedures for answering these questions can probably be found using the indexes of standard texts, or by searching the literature using accepted key words and thesauri.

However, the links between a biological problem and the existing histochemical methods may be less straightforward. When biologists think their problems might be solved by histochemical procedures, the presenting questions are most commonly expressed as:

(1) What is the nature of this object, which I have seen when looking down my microscope?, or
(2) How do I detect — and perhaps assay — entity X in this biological specimen?

So the initial task is to see if histochemical methods can be found in the literature which will, for a particular situation, answer these questions. The first of these questions is less formidably open-ended than it may seem at first sight. After all you will often have a great deal of information concerning the biological context. In particular you will probably know something about the organism or cell containing the unknown object: whether it is healthy or pathological, normal or experimental. Consequently you can usually speculate as to the nature of the unknown object: 'It could be an A, a B or a C'. Hence the problem resolves into searching the histochemical literature for methods to demonstrate entities A, B and C. If you are quite unable to generate such hypotheses, then an initial application of general histochemical test may assist formulation of sharper questions. This unknown cytoplasmic granules may be histochemically tested for the presence of, say, DNA and GAGs. The presence of DNA may suggest an aggregation of virus particles, whilst GAGs could indicate a lysosomal nature. The second of the above questions may also resolve into a search for, and use of, histochemical procedures which demonstrate entity X. There are many compilations of staining procedures which can form the starting point of such literature searches, (see section 9.6. for some examples).

Supposing, however, that a literature search reveals no methods for entity X? Assume, for instance, that you are a mycologist investigating fungal infections of commercial mushrooms. You wish to use a microscopical method, and the question arises, 'How can I stain parasitic hyphae without also staining the fungus in which they are embedded?'. Suppose you find that, whilst routine procedures for demonstrating fungi do stain the hyphae, they also stain the mushroom. Moreover your literature search turns up no selective stain for the parasitic species, and your colleagues offer you no useful suggestions. What do you do next?

Perhaps you can find a sympathetic histochemist — with a wide range of interests and technical skills, and a little spare time — and ask her? This chapter offers help to those lacking such a resource.

9.2 DEFINING AND USING HISTOCHEMICAL MARKERS

If you fail to find a mention of the entity of interest in the histochemical literature, you can next ask, 'In what biochemical, physical or physiological ways might this entity differ from the rest of the specimen?' Suppose you wish to demonstrate a particular cell type. Have you any reason to suppose that these cells are relatively rich in GAGs, or in some particular enzyme or antigen? Might they be more or less dense than their surroundings? Could they be physiologically different — perhaps they are phagocytic?

Having listed such possible differences, you may enquire if any of them could be detected histochemically. For instance, most GAGs can be easily dyed; density differences can give rise to selective staining in diffusion rate-controlled methods; and phagocytosis can be detected by exposing live cells to a suitable probe. When differences exist and can be demonstrated, a potential *histochemical marker* for the entity of interest has been found.

Such a putative marker must be carefully scrutinized. For instance are any unusual maneouvres required in the preparation of the specimen, or in the staining procedure? Are special technical skills required, that you lack? Does the method call for expensive reagents or apparatus? Assuming the histochemical marker is acceptable on these criteria, the next step is to try the method out in practice. If the marker is a histochemically familiar entity, a suitable procedure may be sought in the literature as previously described.

9.3 DESIGNING NEW STAINING PROCEDURES

However, it must be acknowledged that even if target entity/background differences may be specified, the histochemical literature may offer no obvious ways of demonstrating them. Although this may seem an impasse, devising new staining methods is not magic. Or, if it is, the following suggestions are guidelines for histochemical magicians!

The first suggestion is to look for possible analogies with known staining methods, or indeed with chemical and biochemical colorimetric tests, chromatographic spray reagents, and so on. The second suggestion is to consider using the 'labelled antibody battering ram'. And the third is to apply the 'dyeing shotgun', aiming the shotgun with the aid of the structure–staining correlation technique.

Making analogies is straightforward, at least when this involves other histochemical procedures. For instance, imagine you wish to demonstrate a hydrolytic enzyme Zase. Even without detailed knowledge one can suggest using an appropriate naphthyl derivative as substrate, and demonstrating Zase using an azodye technique. Similarly if you wish to demonstrate some metal ion, then thinking of the Alizarin Red method for calcium ions suggests the use of a complexing agent to form a water insoluble chelate with the metal. Note that suitable complexing agents might well be sought outside histochemistry, in the literature of analytical or coordination chemistry. This advice can be generalized: there are many sources of useful analogies. Take chromatography for instance. Affinity chromatography offers examples of an amazing variety of selective target–probe interactions; and in addition the visualizing systems devised for thin layer and paper chromatography provide numerous suggestions of possible histochemical application. Other potential sources of analogies, largely unexplored, include aspects of biochemical pharmacology, for instance studies of selective enzyme inhibitors, and of drug-receptor binding sites.

The notion of raising an antiserum against some component of the entity of interest is now so familiar that no discussion is called for here. Many excellent accounts of immunocytochemical procedures, applications, and difficulties may be found in texts such as that of Polak and Noorden (1983). As indicated by my earlier use of the phrase 'the labelled antibody battering ram' this approach, though widely applicable, may prove expensive in terms of time and resources.

Applying a wide range of dyestuffs to a specimen, hoping that one will be a selective stain for the entity of interest, is another familiar approach. However, this strategy has been revitalized by use of structure–staining correlation techniques to guide selection of suitable reagents. Such correlations may be used in two quite different ways. First the numerical structure parameters describing a not-so-bad reagent can be taken as a starting point. Reagents differing from the starting compound in various ways (bigger, smaller, more or less hydrophilic, and so on) can then be selected, and investigated. After seeing how reagents perform, a second cycle of selection can be made, and so on. The second approach considers the nature of the entity of interest, of the background, and of the proposed staining procedure, and then defines a 'probably effective' staining reagent. Dyes which match the specification are then selected. For accounts of the structure–staining approach, and for some examples of its application, see the papers of Horobin (1981, 1982), Horobin and Fleming (1980), and Jurranz *et al.* (1986).

9.4 CASE EXAMPLES OF SELECTION AND DESIGN

9.4.1 Selection of standard stains, to solve a novel problem

Let us return to the mycologist, seeking a selective stain for the hyphae of a parasitic fungus growing within edible mushrooms. What differences might there be between the two fungi? Will the cell walls differ? Certainly many fungal surfaces carry antigenic determinants; and if the hyphae were strongly antigenic there is no particular reason why the antigens would be the same as those on mushroom cell surfaces. There may also be cruder chemical differences. Might the cytoplasmic contents of the parasitic hyphae differ from the surrounding host cells? Quite probably: after all the hyphae are penetrating rapidly into the host. Consequently the parasite may have a higher rate of protein synthesis, and so more ribosomes; also penetration may be aided by secretion of lysosomal hydrolytic enzymes.

Might any of these, speculative, differences be detectable by histochemical methods, and so be used as histochemical markers? In fact standard methods exist for detecting antigenic materials, for ribosomal RNA, and for hydrolytic enzymes. Some details of such methods, with their advantages and snags, is given in Table 9.1. This summary indicates that the best strategy for the mycologist would be first to try the relatively cheap and simple dyeing methods for RNA. If these do not prove satisfactory, then try the somewhat more complex enzyme histochemical demonstration of hydrolases. Only if these, too, fail to provide selective staining of hyphae should serious thought be given to the more complex and expensive labelled antibody strategy.

9.4.2 Design of a novel stain, to solve an old problem more effectively

As a second example, consider a different kind of question, and a different approach to finding a solution.

Imagine now that a botanist, who routinely uses microscopic methods, wishes to determine the site of deposition of callose, a plant polysaccharide often produced following injury. Although several published methods exist for staining callose, from the botanist's point of view they all have snags. Several use reagents not available commercially; and the botanist does not wish to use methods other workers may have difficulty in repeating. Moreover the one commercially available stain is a fluorochrome, and the botanist does not possess a fluorescence microscope.

The four reagents previously used as callose stains were all dyestuffs, suggesting that a 'shotgun' strategy using a wide variety of dyes may be worth trying. Of course the shotgun can be aimed, using the structure–staining correlation approach. The previously used dyes were all anionic, and three of the four were hydrophilic and possessed large planar conjugated systems. These three features can be modelled numerically using the structure parameters electric charge, log P and largest conjugated fragment values respectively. By taking the numerical values of the parameters of the four previously used dyes, a speculative structural specification of a selective callose stain can be suggested. This numerical specification can then be used to check if any 'maybe callose stains' exist amongst dyes sold by laboratory suppliers. If such dyes are found, they can be purchased and tested.

9.5 SELECTING AND DESIGNING — AN OVERVIEW

Since the approach in this chapter may be unfamiliar, a graphic summary is provided in Fig. 9.1.

Table 9.1 — Methods for staining possible histochemical markers of parasitic fungi

Possible histochemical markers	Possible staining methods	Technical problems anticipated	Expected validity of staining	Resources required
Surface antigens	Labelled antibodies	Antibodies not commercially available. Moreover in-house preparation requires a sample of hyphae free of mushroom — possibly a difficulty	High selectivity, assuming problem overcome. Use of peroxidase label would allow formalin fixation, hence good morphology	Considerable efforts by an immunologically competent worker needed
Ribosomal RNA	Methyl Green — Pyronin or Cuprolinic Blue dyeing methods	Pure Pyronin needed which is commercially available	Good, especially if Cuprolinic Blue method used	Simple and cheap
Hydrolytic enzymes, e.g. acid phosphatase or peptidase	Metal ion method for phosphatase; azo dye methods for both	Glycolmethacrylate embedding may need extended incubation times	High selectivity. Localization best if the plastic embedding is used	Procedures are standard, and not very costly

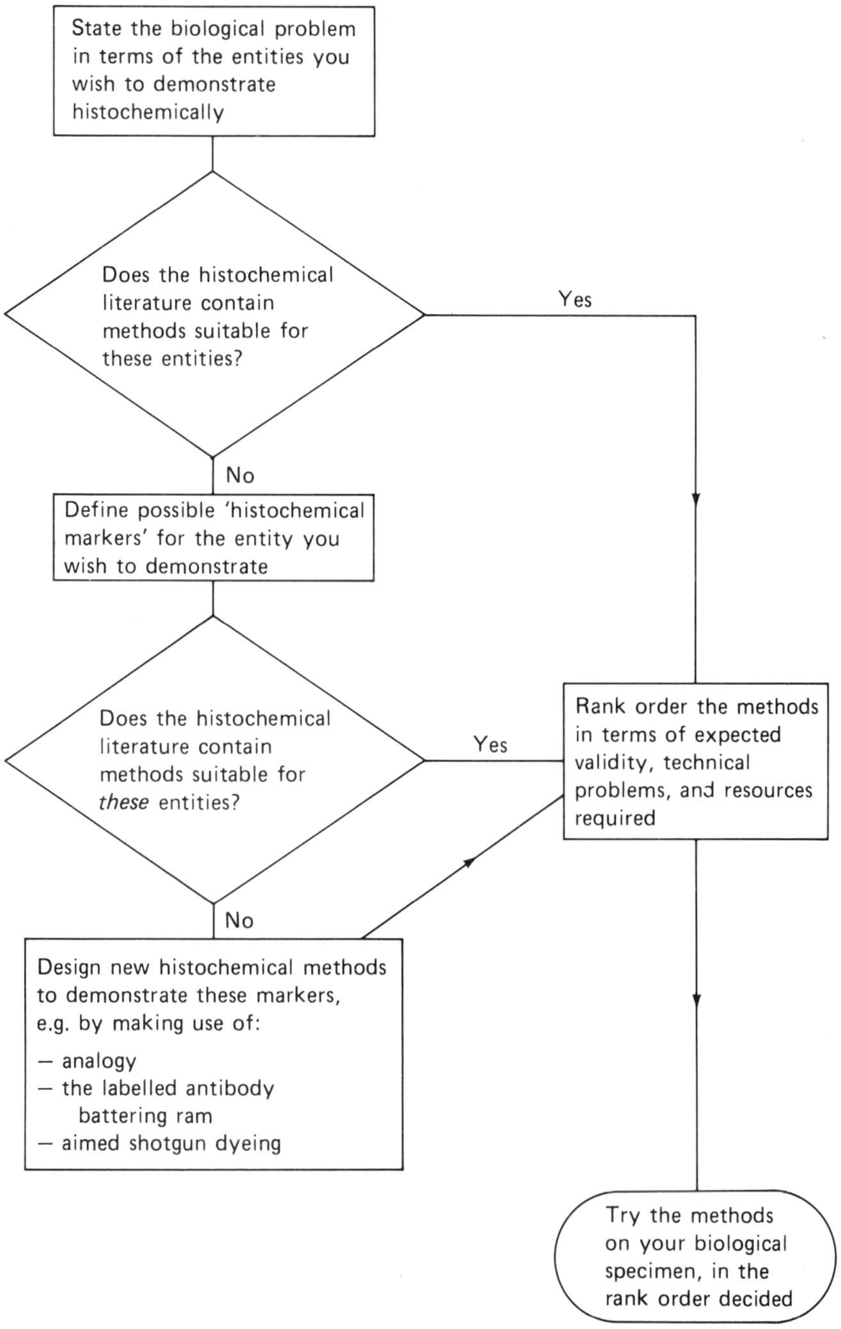

Fig. 9.1 — A simple algorithm illustrating the selection and design of histochemical staining methods.

9.6 FURTHER READING

Some histochemical texts are arranged to facilitate choosing methods for visualizing various types of chemical entity (Chayen and Bitensky, 1973; Gahan, 1984). However, I am not aware of extended attempts to connect the questions asked by biologists to appropriate staining methods, nor of accounts discussing non-chemical entities. Even descriptions of the process of designing novel procedures — as distinct from descriptions of the procedures themselves — are uncommon. For a recent example where this process *is* made explicit see Scott (1980).

If you are looking for existing methods to demonstrate a variety of named entities, Table 9.2 provides a number of starting points.

Table 9.2 — Some information sources on the preparation and staining of biological specimens

Topic	Sources
Specimen preparation methods	
Fixation, embedding, mounting, and so on	Bancroft and Stevens (1982), Lillie and Fulmer (1976), Pearse (1968/1972, etc.), and journal *Stain Technology*
Possible artifacts	Thompson and Luna (1978)
Type of specimen	
Bacteriological	Bancroft and Stevens (1982), Clark (1981), Lillie and Fulmer (1976)
Botanical	Clark (1981), Gahan (1984), Jensen (1962)
General	Clark (1981), Lillie and Fulmer (1976), Pearse (1968/1972, etc.), Thompson (1966), and the journals *Histochemistry, Histochemical Journal, Journal of Histochemistry & Cytochemistry* and *Stain Technology*
Histopathological (i.e. human)	Bancroft and Stevens (1982), Lillie and Fulmer (1976), Pearse (1968/1972, etc.), Thompson (1966)
Invertebrate	Clark (1981)
Mycological	Bancroft and Stevens (1982), Clark (1981), Lillie and Fulmer (1976)
Neurological	Adams (1965), Bancroft and Stevens (1982), Lillie and Fulmer (1976)
Protozoan	Clark (1981), Lillie and Fulmer (1976)

Appendices

Appendix 1 Physicochemical characteristics of staining reagents

The properties and parameters tabulated below have been found to give insight into many staining mechanisms. Obviously acid and basic dyeing require dyes whose electric charges (*Z*) are negative and positive respectively. Rates of reagent diffusion are influenced by the size of the reagent; which feature can usually be modelled by the molecular or ionic weight (M/I wt). Van der Waals attractions, those universal contributions to reagent–tissue and reagent–reagent affinities, tend to increase with increases in the value of the conjugated bond number (CBN). The contribution of hydrophobic bonding to reagent–tissue affinity usually increases with increasing hydrophobicity of the reagent, a convenient measure of which is the logarithm of octanol–water partition coefficient (log *P*), an improved version of the Hansch π parameter.

Some simple applications of the tabulated information follow. In section 3.3.3 the high affinities of bisazo dyes and bisformazan pigments for tissues were contrasted with the lower affinities of monoazo and monoformazan analogues. These differences were linked to increased van der Waals forces occurring with the larger compounds. The sizes of the conjugated systems can be modelled by the conjugated bond numbers. This bisazo and monoazo dyes Congo Red and Ponceau 2R have CBNs of 43 and 21 respectively, whilst the analogous Nitro Blue and Iodo Formazans have CBNs of 57 and 28. In section 8.1.2 a general structure–staining correlation was made concerning one-bath Trichrome stains, namely, when staining is carried out with a low pH, aqueous solution of two acid dyes, selective colouration of collagen fibres with the larger dye occurs when the anionic weights of the dyes differ by more than 200 daltons. As final example consider the hydrophobic lipid stains discussed in section 8.1.5. Again a general structure–staining correlation can be specified: effective lipid staining requires the dye to have a log *P* value >5.0; Nile Red, Oil Red O, and Sudan Black B, three currently favoured lipid stains, have log *P* values of +5.1, +9.0 and +7.2 respectively.

The tabulated values are not experimental measures, but are obtained by inspection of structural formulae (e.g. electric charge), or by simple pencil-and-paper estimates (e.g. molecular weight). For details and examples of such calculations see, for instance, section 11.14 of my book (Horobin, 1982).

Physicochemical characteristics of staining reagents

Name of reagent	C.I. no.	*Z*	M/I wt	CBN	log *P*
Acid Fuchsin	42685	−1	540	24	−13.5
Alcec Blue		+3	1017	48	−11.1
Alcian Blue 8G	74240	+4	1380	48	− 9.4
Aldehyde Fuchsin		+1	668	27	+ 4.0
Alizarin Brilliant Blue	63010	−1	349	22	− 4.9
Alizarin Red S	58005	−1	319	20	− 2.8
Aluminium Haematoxylin		+1	629	36	−ca.7
Astra Blue 6BLL		+3	972	48	−10.4
Azur A	52005	+1	256	18	− 1.5
Azur B	52010	+1	269	18	− 1.1
Carminic Acid	75470	−1	492	27	− 2.0
Chlorantine Fast Green	34045	−5	1218	67	−11.8
Congo Red	22120	−2	651	43	− 1.9
Cresyl Fast Violet		+1	340	23	− 0.7
Crystal Violet	42555	+1	372	24	+ 1.2
Cuprolinic Blue		+4	640	44	−13.4
Direct Blue 152	24366	−2	742	47	− 3.3
Eosin Y	45380	−2	646	31	+ 1.4
Evans Blue	23860	−4	869	45	− 8.7
Fluorescein	45350	−2	330	28	− 2.4
Gallein	45445	−2	361	36	− 3.8
Gallocyanin Chrome Alum		+1	724	44	−ca. 6.5
Hematein, anionic		−1	299	18	− 2.8
Hoechst 33258	33258	+1	425	30	− 1.2
Iodo Formazan		0	471	29	+ 6.2
Light Green SF	42095	−2	747	38	− 6.6
Luxol Fast Blue FBG	34200	−4	965	56	− 6.9
Methylene Blue	52015	+1	285	18	− 0.9
Methyl Green	42585	+2	387	23	− 1.2
Neutral Red	50040	+1	254	18	− 1.0
Nigrosin, spirit soluble	50415B	+3	797	71	+ 1.3
Nile Red		0	317	23	+ 5.1
Nitro Blue Formazan		0	734	57	+10.0
Nitro Red		−2	466	25	− 6.5
Oil Red O	26125	0	409	30	+ 9.0
Orange G	16230	−2	406	21	− 5.0
Orcein		+1	585	36	+ 1.0
Pararosanilin	42500	+1	288	24	− 1.5
Picric Acid	10305	−1	228	17	+ 0.9
Pyronin Y	45005	+1	257	18	− 0.2

Name of reagent	C.I. no.	*Z*	M/I wt	CBN	log *P*
Quinacrine		+1	437	19	+ 0.1
Quinacrine mustard		+1	471	19	− 0.1
Safranin O	50240	+1	315	25	+ 0.4
Sirius Red F3B	35780	−6	1253	62	−17.1
Sudan Black B	26150	0	456	35	+ 7.2
Ponceau 2R	16150	−2	434	21	− 3.3
Tetranitro Blue (TNBT) Formazan		0	824	63	+ 9.4
Thiazolyl (MTT) Formazan		0	335	21	+ 4.9
Thiazolyl–cobalt (MTT–Co) Formazan		0	725	42	+ 8.4
Thionin	52000	+1	225	18	− 2.1
Toluidine Blue	52070	+1	270	18	− 0.9

Appendix 2 A glossary of some histochemical terms

Note: cross-references within the glossary are indicated by use of **bold** typeface.

Acid dye. A dye whose coloured component is an **anion**, for instance Congo Red and Eosin Y. Such dyes are *not* acids, the term derives from traditional textile dyeing usage where they are used to colour protein fibres from acid dyebaths.

Acidophilic. Literally, acid [dye] lover: tissue components, typically proteins, which take up **acid dyes**. Examples of acidophilic tissue components are collagen fibres and erythrocytes.

Affinity. The tendency of a staining reagent (e.g. **dye** or **labelled antibody** or **enzyme substrate**) to move from the staining solution onto the tissue. *Contributions to affinity* are factors favouring this, such as **coulombic forces** and other reagent–tissue attractions, or positive entropy changes, for instance **hydrophobic bonding**. An alternative usage defines affinity as those attractive forces binding stains to the tissue, which from the perspective of this present book is merely one of the contributions to affinity.

Aggregation of dyes. **Dye** molecules do tend to stick onto things, including other dye molecules. The resulting dye aggregates may contain two molecules or a hundred. Aggregation occurs both in solution and in stained tissues (see **metachromasia**). Structural features favouring aggregation include large **conjugated bond numbers** (facilitating dye–dye **van der Waals attractions**) and, if staining solutions are aqueous, a **hydrophobic** character (facilitating dye–dye **hydrophobic bonding**).

Amplification of staining. Use of a secondary staining to increase the intensity of colouration of structures stained by a previous procedure, without alteration of the initial pattern of **selectivity**. Thus gold toning can be used to amplify an initial

silver stain; and a stain comprised of oxidized **diaminobenzidine**, derived from a peroxidase **labelled antibody**, can be amplified by silver staining.

Aniline dye. Anachronistic term for **synthetic dye**, originating from the time in the nineteenth century when aniline was the most common precursor of synthetic dyes.

Anion. An ionic species carrying a negative charge, for example an **acid dye** or a chloride ion.

Artifact. 1. An entity generated by *Homo sapiens.* 2. All stained specimens are artifacts in the above sense, and the histochemical usage of the term implies an informationally misleading artifact, resulting from technical error, or from lack of understanding of the specimen or the preparation and staining processes.

Azo dye. A **dye** containing one (monoazo), two (bisazo) or more (polyazo) azo group(s); that is –N=N–.

Basic dye. A **dye** whose coloured component is a **cation**, for instance Alcian Blue, Aluminium Haematoxylin, and Neutral Red. Such dyes are *not* bases, the term derives from textile dyeing usage, in which they were once used to colour protein fibres from alkaline ('basic') dyebaths.

Basophilic. Literally, basic [dye] lover: namely tissue components which take up basic dyes. For instance nuclear chromatin, and mast cell granules.

Best's carmine. A traditional selective stain for glycogen, containing the **natural dye** Carminic Acid, dissolved in an alkaline, salty aqueous–methanolic solution. Surprisingly, all components of Dr Best's stain prove necessary for effective staining.

Biopolymer. General term for those high molecular weight compounds characteristic of living organisms: nucleic acids, polysaccharides, proteins and so on.

Blockade. Conversion of a tissue site capable of binding or generating stain into a non-binding, non-generating form. Examples are: binding unlabelled antibodies to tissue antigenic sites, to block access of labelled immunoglobulins; inhibition of enzymes by treatment with glutaraldehyde; and esterification of tissue.

Cation. An ionic species carrying a positive charge, such as a **basic dye** or a sodium ion.

Colloid stabilizer. 1. Water soluble, high molecular weight compound (for example, agar or polyvinylalcohol) added to incubation solutions in **enzyme histochemistry** to prevent losses of **biopolymers** from the cells and tissues. 2. Occasionally used as a synonym for the protection colloids found in physical developers.

Conjugated bonds. A linked set of **covalent bonds** in which it is not possible to spatially localize all the bonding electrons. For example, the benzene carbon skeleton is often drawn as if made up of three carbon–carbon single bonds plus three carbon–carbon double bonds. In fact, since the double bonds are *conjugated,* the pi-electrons are delocalized, and all six bonds are identical. Such delocalized electrons are very mobile, so electrical influences are easily propagated from one part a conjugated molecule to another, enhancing **dipoles** and **polarizability**, so favouring **van der Waals forces.**

Conjugated bond number. A numerical **structure parameter**, describing the overall extent of **conjugation** within a stain or staining reagent. Usual abbreviation: CBN.

Coulombic forces. Electrical attractions and repulsions, resulting from the existence of positively and negatively charged species, such as $-NH_3^+$ and $-SO_3^-$.

Covalent bond. A link between two atoms resulting from their sharing, usually, a pair of electrons (as in a C–C bond). In double bonds two pairs of electrons are shared (for instance C=O and C=C). Though drawn as equal, the two bonds are somewhat different in character, and in one of them the electrons are much more loosely held between the two atoms than in the other.

Crosslinking. Links between polymeric chains or, less often, lower molecular weight molecules. 1. Generated by a fixative agent, for example crosslinking of proteins by glutaraldehyde, or of unsaturated lipids by osmium tetroxide. 2. Present in native proteins, commonly in the form of disulphide bridges between or within polypeptide chains. 3. Present after polymerization in some **plastic embedding media**.

Decalcification. Removal of hard deposits of calcium salts from tissues, prior to microtomy. Decalcification agents include acids, and the calcium chelating agent ethylenediamine tetraacetic acid.

Denaturants. Agents which cause **denaturation** of proteins. For instance intangibles, such as heat, organic solvents, and chemically reactive compounds, such as formaldehyde or dichromates.

Denaturation. The destruction of the native structural form of a protein, resulting in an insoluble, typically more **hydrophobic**, molecule. Typically this reduces the antigenicity, and enzymicity when present, of the native protein. Denaturation may be effected by a wide variety of **denaturants**.

Diaminobenzidine. A slightly water soluble, polyamino primary aromatic amine. Its oxidation, which can be catalysed by the enzyme peroxidase, yields various brown-black polymeric **pigments**.

Differentiation. Histochemical usage: a process of controlled de-staining.

Dimethylglyoxime. A chelating agent, which yields brightly coloured, insoluble **metal coordination complexes** with nickel and several other metal ions.

Dipole. The accumulation of negative charge on one atom of a covalently linked pair, the other atom having a partial positive charge. The partial negative charge is acquired by the most **electronegative** atom. Thus in a nitro group the oxygen atoms are more electronegative than the nitrogen, and so carry partial negative charges. The *dipole moment* of a dipole is the product of the magnitude of the charge on the electonegative atom, and the distance between the electronegative and electropositive atoms.

Dye. A coloured organic ion or molecules, whose ability to absorb light is due to an extensive system of **conjugated bonds.** A solution of a dye is capable of imparting colour to cells and tissues immersed in it, by dye moving from solution into the biological specimen. Such *dyeing* need not involve changes in **covalent bonds** or in oxidation state.

Electronegativity. The ability of an atom to attract an electron. For example chlorine atoms are very electronegative, hence the stability of Cl^-; sodium is only slightly electronegative ('electropositive'), hence the stability of Na^+.

Embedding. Enclosing cells and tissues within a matrix, initially liquid, such as molten paraffin wax or a **plastic** monomer. This is then solidified, for instance by freezing or polymerization respectively.

Enzyme histochemistry. A strategy for demonstrating enzymes in cells and tissues, utilizing the catalytic activities of the undenatured enzymes. Suitable *enzyme substrates* are enzymically converted, directly or indirectly, into coloured, immobile **final reaction products** which mark the enzymic site. When the conversion is indirect, the *intermediate reaction products* are transformed into the final products by reaction with **visualization agents**.

Enzyme substrate. See **enzyme histochemistry**.

Fading. Transformation of coloured stains present in the tissues into colourless derivatives, either due to light (such as sunlight, or the UV of a fluorescence microscope) or to the chemical environment (such as the **mounting medium**).

Final reaction product. The coloured, immobile (owing to adsorption onto tissue proteins, insolubility, or both) terminal reaction product marking the enzymic sites after **enzyme histochemical** staining.

Fixation. The conversion of certain components of cells and tissues into insoluble form. The objectives are the preservation of the morphology, and the chemical reactivity, of the biological materials during processing, staining, and microscopical observation. Mechanistically, fixation may involve **crosslinking, denaturation, trapping**, etc.

Flow cytometry. A system for determining the numbers of cells of different types within a population, for example in peripheral blood. Suspended cells flow in a narrow jet of fluid past a battery of detectors, e.g. measurements are made of absorption, fluorescence, and light scattering.

Fluorescence. The re-emission of light (or other electromagnetic radiation) of slightly longer wavelength than the radiation previously absorbed by the fluorescent molecule.

Fluorochrome. A stain which exhibits **fluorescence** whilst bound to cells and tissues.

Formazan. A **pigment** produced by the oxidation (chemically or enzymically) of a colourless, water-soluble tetrazolium salt [histochemical usage].

Glare. Unfocused light from the object, diffusely spread across the image owing to multiple reflections in the optical system. Glare visually degrades the image. Synonym: stray light.

Hydration. The tight binding of water molecules to molecular groupings, for example, –OH in glycogen, $-NH_3^+$ and $-CO_2^-$ in proteins. Such water (bound by **dipoles** and **hydrogen bonding**) can form a substantial *hydration shell* around a **biopolymer**; thus water can amount to more than half the weight of a sample of protein.

Hydrogen bond. A weak, direct bond between a hydrogen atom and an adjacent electronegative atom (usually oxygen or nitrogen) to which it is not linked by a **covalent bond**.

Hydrophilic. Literally, water loving. A species with an affinity for water (for instance Alcian Blue, glycogen, sodium chloride); also those molecular fragments favouring such affinity (such as $-SO_3^-$ and –CONH–). See also **hydration** and **hydrophobic**.

Hydrophobic. Literally, water hating. A species with little or no affinity for water; also the molecular fragments favouring such distaste (for example $-CH_3$ and $-C_6H_5$). Hydrophobic compounds are often, though not always, **lipophilic**. The hydrophobic character of a dye or other reagent may be assessed using the **structure parameter**, log P (namely the logarithm of the water–octanol partition coefficient).

Hydrophobic bonding. The coming together of hydrophobic molecules or group ings originally dispersed within an aqueous environment. Examples of phenomena driven by hydrophobic bonding include folding of polypeptide chains bringing together hydrophobic amino acid residues in the core of the protein; and the staining of a hydrophobic biopolymer by a hydrophobic dye. Although termed hydrophobic *bonding*, the process is driven by the increasing **entropy** of the aqueous solvent, not by attractive forces between the hydrophobic groups.

Immunostaining. Histochemical staining generated by the use of **labelled antibodies**, q.v.

Intermediate reaction product. See **enzyme histochemistry**.

Ionic crystal. A regular, three-dimensional array of anions and cations (e.g. lead sulphide is Pb^{2+} plus S^{2-}) whose overall charge is zero.

Ionic weight. The molecular weight of a single ion, ignoring the counterion. Thus the **acid dye** Eosin Y could be obtained as the potassium or sodium salts, whose molecular weights differ. However, the ionic weight of the coloured Eosin anion is the same in both salts.

Labelled antibodies. A widely used class of staining reagents. The antibody component provides the **affinity** and **selectivity** for binding the reagents to tissue antigens. The label takes one of two forms. Some give microscopic images directly, for instance the **fluorochrome** FITC and colloidal gold. Other labels are themselves invisible but can give rise to visible products, for example peroxidase, which can be demonstrated by using **enzyme histochemistry**.

Lipophilic. A compound soluble in lipid, for instance a **sudan dye**; or a solvent of lipids, such as chloroform or xylene. Literally, fat-loving.

Masking. The inhibition of the staining of one tissue component by a second tissue component.

Metachromasia. A phenomenon in which a single dye gives rise to two or more different colours when bound to different tissue components. *Metachromatic dyes* exhibiting this effect include Methylene Blue and **fluorochromes** such as Acridine Orange. Mechanisms are varied, but often involve tissue-mediated dye **aggregation**.

Metal coordination complex. Metal ions to which are bound, by polar **covalent bonds**, a number of ligands. Complexes of histochemical interest include (a) the anionic component of **Prussian Blue**, an inorganic **pigment**, in which the ligands are CN^- and the metal cation Fe^{2+}; and (b) the **mordant dye** Gallocyanin Chrome Alum, in which the metal ion is Cr^{3+} and the ligands two molecules of the dye Gallocyanin plus two water molecules.

Mordant dye. 1. A dye capable of forming a metal coordination complex; or such a complex (for example Hematein or Gallocyanin Chrome Alum respectively). 2. A dye supposed to be bound to the tissues via dye–metal ion, metal ion–tissue bonds.

Natural dye. A dye synthesized by a plant (such as *Haematoxylon campechianum*), a beetle (for instance *Coccus cacti*), or other non-human creatures; compare **synthetic dye**.

Negative stain. A stain which is deposited against the surface of a cell or tissue, hence displaying its morphology; the entity of interest is not stained.

Neutral. An ionic dye in which both anion and cation are dyestuffs. For example the Romanowsky stains, the active component of which is an Azure B eosinate. A neutral dye is thus a salt, and not non-ionic.

Non-ionic dye. A **dye** in which the coloured entities are not ionic. Some are **lipophilic**; some (for instance Hematein and Purpurin) are not.

Pigment. A coloured, insoluble substance. It may be either organic (such as a **formazan**) or inorganic (for example **Prussian Blue**).

Plastic embedding media. Embedding media generated by infiltrating tissues with liquid monomers, which are then polymerized *in situ.* Such media often give improved morphological preservation, and also facilitate the microtomy of hard structures. Some plastic media (such as methyl methacrylate) are removed from the tissue prior to staining; others (for instance glycol methacrylate, epoxy resins) are not. All require changes in those staining schedules devised for paraffin or cryostat sections, to a greater or lesser degree. The term *plastic sections* is commonly used to describe sections cut from specimens embedded in such plastic media.

Polarizability. A measure of how easy it is to induce a **dipole moment** in a molecule or grouping, by applying an electric field (for example due to the **dipole** of another molecule or grouping).

Polyanions. Polymeric molecules whose repeat units carry anionic charges. For example DNA (the anionic groups being phosphates) and GAGs (the anionic groups being carboxylate and/or sulphate).

Polychromasia. Synonyms: polychrome or polychromatic staining. Staining of different structures in a cell or tissue different colours, by one stain or several. Thus **Romanowsky dyes** give rise to polychromasia in smears of peripheral blood.

Progressive staining. A strategy of selective colouration in which an initially colourless tissue preparation is stained until the required amount of stain is taken up. Compare **regressive staining**.

Prussian Blue. An inorganic **pigment**, arising as a reaction product in the Perls procedure for the demonstration of Fe^{3+} in the tissues; Prussian Blue is the ferric salt of an anionic **metal coordination complex**, $Fe(CN)_6^{4-}$.

Quenching. 1. Of fluorescence: the processes whereby stains which might be expected to fluoresce do not, owing to their electronic energy being transformed (for example into heat) or transferred (for instance to solvent, tissue, other dyes). 2. Of tissues: rapid cooling in a suitable medium (such as liquid nitrogen) prior to cryotomy.

Regressive staining. A strategy of selective staining, involving the **differentiation** of an overstained tissue preparation, Compare **progressive staining**.

Romanowsky dyes. Traditionally, these are complex mixtures of Eosin plus Methylene Blue and its lower homologues, devised in the nineteenth century by a Russian haematologist to give **polychrome staining** of blood cells. It is now known that Azur B and Eosin are the only necessary components. The *Romanowsky–Giemsa effect* refers to the rich purple staining of nuclear chromatin and certain cytoplasmic granules, contrasting with the blue staining of lymphocyte and blast cell cytoplasms, and the red staining of eosinophil granules and erythrocytes.

Selectivity of staining. The intensity of staining of the target structure or substance compared with that of the background.

Sensitivity of staining. The ability, or otherwise, of a staining procedure to detect small amounts of a tissue component. Thus **fluorochromes** are often much more sensitive than otherwise similar dyes.

Silver staining. This involves the transformation of the Ag^+ present in the staining solution into microcrystals of metallic silver in the tissues, giving rise to colours in the range yellow-brown-black. Staining mechanisms vary from direct reduction of Ag^+ by tissue constitutents, to catalysis by the tissue of the reaction between the silver ions and reducing agent present in the staining reagent.

Specificity of staining. The ability of a staining procedure to colour a particular tissue component, and no other. There are few specific stains, fewer than most believe. Even **labelled antibodies** readily give positive artifactual staining, owing to non-immunological effects.

Spinner. A mechanical device used by haematologists for preparing uniform layers of peripheral blood cells.

Staining. The visual labelling of an entity (which may be anything from a molecular fragment to a cellular process) by attaching or depositing in its vicinity a marker of characteristic colour or shape.

Structure parameter. A number used to describe or model a structural/physicochemical feature of a stain or staining reagent. For instance the sizes of **non-ionic dyes** can be modelled by their molecular weights, the extent of their conjugated-systems by their **conjugated bond numbers**, and their **hydrophobic** or **hydrophilic** nature by their log *P* values.

Substrate. See **enzyme substrate**.

Sudan dye. Non-ionic, **lipophilic** dye. Term derives from the commercial names of many excellent lipophilic dyes, which take the form 'Sudan X' (where X=III, Black, Red, etc.).

Synthetic dye. A dye synthesized by *Homo sapiens*, typically from precursors derived from crude oil. Compare *natural dye.*

Theory. A useful story for guiding and justifying practical actions. Scientific theories are preferably predictive, couched in suitably sober language, and capable of disproof.

Trapping. 1. A mechanism of fixation, in which soluble compounds of unaltered nature (such as glycogen, insulin) are caught within a cage of insolubilized proteins. 2. The process whereby a soluble, freely diffusing **intermediate reaction product** is transformed by reaction with a *trapping agent* into an insoluble derivative, and so retained within the tissue.

Van der Waals forces. Term describing various non-directed, short-range, polar forces (e.g. dipole–dipole, dipole–induced dipole, and dispersion forces) which bind molecules together.

Visualizing agent. A reagent used in **enzyme histochemistry** to convert a colourless **intermediate reaction product** into a coloured final **reaction product**. For instance, when demonstrating phosphatases, ammonium sulphide is often used to convert colourless lead phosphate into black lead sulphide.

Vital staining. The process whereby a **stain** is taken into a live organism or cell, sometimes passively, sometimes by an active transport mechanism. In any event, the distribution of stain reflects the physiological activities of the cell or creature.

References

Adams, C. W. M. (1965) *Neurochemistry*. Amsterdam, Elsevier.

Baker, J. R. (1958) *Principles of biological microtechnique: a study of fixation and dyeing*. London, Methuen.

Baker, J. R. (1962) Experiments on the action of mordants. II. aluminium hematein. *Quart. J. micr. Sci.* **103** 493–517.

Bancroft, J. D. & Stevens, A. (1982) *Theory and practice of histological techniques*, 2nd edn. Edinburgh, Churchill-Livingstone.

Bartholomew, J. W. (1962) Variables influencing results, and the precise definition of steps in Gram staining as a means of standardising the results obtained. *Stain Technol.* **37** 139–155.

Bauman, J. G. J., Wiegant, J. & van Duijn, P. (1981a and b) Cytochemical hybridisation of fluorochrome-labelled RNA. 1. Development of a method using nucleic acids bound to agarose beads as a model. 2. Applications. *J. Histochem. Cytochem.* **29** 227–237 and 238–246.

Bauman, J. G. J., van der Ploeg, M. & van Duijn, P. (1984) Fluorescent hybridocytochemical procedures. In *Investigative microtechniques in medicine and biology* (eds. Chayen, J. and Bitensky, L.). New York, Dekker, pp. 41–87.

Bayliss High, O. B. (1982) Lipids. In *Theory and practice of histological techniques*, 2nd edn. (eds Bancroft, J. D. and Stevens, A.). Edinburgh, Churchill-Livingstone, pp. 217–241.

Bennion, P. J. & Horobin, R. W. (1974) Some effects of salts on staining: use of the Donnan equilibrium to describe staining of tissue sections with acid and basic dyes. *Histochemistry* **39** 71–82.

Boon, M. E. & Druijver, J. S. (1986) *Routine cytological staining techniques*. London, Macmillan.

Boon, M. E. & Kok, L. P. (eds) (1986) *Standardisation and quantitation of diagnostic staining in cytology*. Leiden, Coulomb Press.

Caspersson, T., Zech, L. & Johanson, C. (1970) Differential banding of alkylating fluorochromes in human chromosomes. *Exp. Cell Research.* **60** 315–319.

Chayen, J. & Bitensky, L. (1973 *Practical Histochemistry*, 2nd edn. Chichester, John Wiley.

Clark, G. (1981) *Staining procedures,* 4th edn. Baltimore, Williams & Wilkins.

Clasen, R., Simon, R. G., Scott, R. Pandolfi, S., Laing, I. & Lesak, A. (1973) The staining of the myelin sheath by Luxol dye techniques. *J. Neuropath. expt. Neurol.* **32** 271–283.

Cook, H. (1982) Carbohydrates. In *Theory and practice of histological techniques*, 2nd edn. (eds Bancroft, J. & Stevens, A.). Edinburgh, Churchill-Livingstone, pp. 180–216.

Curtis, D. & Horobin, R. W. (1982) Chromosome banding: specification of structural features of dyes giving rise to G-banding. *Histochemical J.* **143** 911–928.

Elleder, M. & Lojda, Z. (1972) Studies in lipid histochemistry. IX. The specificity of the Holczinger's reaction for fatty acids. *Histochemie* **32** 301–305.

Gahan, P. B. (1984) *Plant histochemistry and cytochemistry: an introduction.* New York, Academic Press.

Gallyas, F. (1979) Factors affecting the formation of metallic silver and the binding of silver ions by the tissue components. *Histochemistry* **64** 97–109.

Goldstein, D. J. (1961) Mechanism of differential staining of nucleic acids. *Nature (Lond.)* **191** 407–408.

Goldstein, D. J. (1963) An approach to the thermodynamics of histological dyeing, illustrated by experiments with Azur A. *Quart. J. micr. Sci.* **104** 424–439.

Goldstein, D. J. (1965) Relation of effective thickness and refractive index to permeability of tissue components in fixed sections. *J. roy. micr. Soc.* **84** 43–54.

Goldstein, D. J. & Horobin, R. W. (1974) Rate factors in staining by Alcian Blue. *Histochem. J.* **6** 157–174.

Hanker, J. S., Kasler, F., Bloom, M. G., Copeland, J. S. & Seligman, A. M. (1967) Coordination polymers of osmium: the nature of osmium black. *Science* **156** 1737–1738.

Hayat, M. A. (1981) *Fixation for electron microscopy.* New York, Academic Press.

Holczinger, L. (1959) Histochemischer Nachweis freier Fettsauren. *Acta Histochem.* **8** 167–175.

Hopwood, D. (1985) Cell and tissue fixation, 1962–1982. *Histochem. J.* **17** 389–442.

Horobin, R. W. (1981) Structure–staining relationships in histochemistry and biological staining. 3. Some comments on the intentional and artifactual staining of lipids. *Acta Histochem.*, suppl. **24** 237–246.

Horobin, R. W. (1982) Selection of optimum tetrazolium salts for use in histochemistry: the value of structure–staining correlations. *Histochem. J.* **14** 301–310.

Horobin, R. W. (1982) *Histochemistry: an explanatory outline of histochemistry and biophysical staining.* Stuttgart, Fischer; London, Butterworths.

Horobin, R. W. (1983) Staining plastic sections: a review of problems, explanations and possible solutions. *J. Microsc.* **31** 173–186.

Horobin, R. W. & Flemming, L. (1980) Structure–staining relationships in histochemistry and biological staining. 2. Mechanistic and practical aspects of the staining of elastic fibres. *J. Microsc.* **119** 357–372.

Horobin, R. W. & Flemming, L. (1988) One-bath trichrome staining — investigation of a general mechanism, based on a structure–staining correlation analysis. *Histochemical Journal.* **20** 35–40.

Horobin, R. W. & Goldstein, D. J. (1974) The influence of salt on the staining of tissue sections with basic dyes: an investigation into the general applicability of the critical electrolyte concentration theory. *Histochem. J.* **6** 599–609.

Horobin, R. W. & Kevill-Davies, I. (1971) A mechanistic study of the histochemical reactions between aldehydes and Basic Fuchsin and acid alcohol used as a simplified substitute for Schiff's reagent. *Histochem. J.* **3** 371–378.

Horobin, R. W. & Murgatroyd, L. B. (1971) The staining of glycogen with Best's Carmine and similar hydrogen bonding dyes. A mechanistic study. *Histochemical J.* **3** 1–9.

Horobin, R. W. & Walters, K. (1987) Understanding Romanowsky staining. I. The Romanowsky–Giemsa effect in blood smears. *Histochemistry* **86** 331–336.

Hoyer, P. E., Lyon, H., Jakobsen, P. & Andersen, A. P. (1986) Standardized Methyl Green–Pyronin Y procedures using pure dyes. *Histochem. J.* **18** 90–94.

ICSH (1984) ICSH Reference Method for staining of blood and marrow films by azur B and eosin Y (Romanowsky–Giemsa stain). *Br. J. Haematol.* **57** 707–710.

Jensen, W. A. (1962) *Botanical histochemistry, principles and practice.* San Francisco, Freeman.

Juarranz, A., Horobin, R. W. & Proctor, G. B. (1986) Prediction of *in situ* fluorescence of histochemical reagents using a structure–staining correlation procedure. *Histochemistry* **84** 426–431.

Kasten, F. H. (1960) The chemistry of Schiff's reagent. *Int. rev. Cytol.* **10** 1–100.

Kasten, F. H. (1971) Cytophotometric investigations on the influence of fixation on Feulgen hydrolysis: evidence for a random loss of purines and pyrimidines from DNA during acid hydrolysis. *Acta histochem.* suppl. **9** 637–647.

Kiernan, J. A. (1981) *Histological and histochemical methods: theory and practice.* Oxford, Pergamon.

Lillie, R. D. (1977) *Conn's biological stains*, 9th edn. Baltimore, Williams & Wilkins.

Lillie, R. D. and Fulmer, H. M. (1976) *Histopathologic technic and practical histochemistry*, 4th edn. New York, McGraw-Hill.

Litwin, J. A. (1985) Light microscopic histochemistry on plastic sections. *Prog. Histochem. Cytochem.* **16**, no. 2. Fischer, Stuttgart.

Macgregor, H. C. & Varley, J. M. (1983) *Working with animal chromosomes.* Chichester, John Wiley.

McGee-Russell, S. M. (1958) Histochemical methods for calcium. *J. Histochem. Cytochem.* **6** 22–42.

Marcus, Y. & Kertes, A. S. (1969) *Ion-exchange and solvent extraction of metal complexes.* London, John Wiley.

Marshall, P. N. & Horobin, R. W. The chemical nature of the gallocyanin–chrome alum staining complex. *Stain Technol.* **47** 155–161.

Marshall, P. N. & Horobin, R. W. (1973a) The mechanism of action of 'mordant' dyes — a study using preformed metal complexes. *Histochemie* **35** 261–371.

Marshall, P. N. & Horobin, R. W. (1973b) The influence of inorganic salts when staining with preformed metal complex dyes. *Histochemie* **37** 299–311.

Masaro, G. D. & Davis, L. D. (1984) Demonstration of β-adrenergic receptors in rat bronchiolar epithelial cells employing 9-aminoacridyl Propranolol fluorescent microscopy. *J. Histochem. Cytochem.* **32** 122–123.

Mendelson, D., Tas, J. & James, J. (1983) Cuprolinic Blue: a specific dye for single-stranded RNA in the presence of magnesium chloride. II. Practical applications for light microscopy. *Histochemical J.* **15** 1113–1121.

Nettleton, G. S. & Carpenter, A. M. (1977) Studies of the mechanism of the periodate acid–Schiff histochemical reaction for glycogen using infra-red spectroscopy and model chemical compounds. *Stain Technol.* **52** 63–77.

Ovadia, M. & Stoward, P. J. (1971) Some quantitative histochemical studies of the periodic acid oxidation of glycogen in situ. *Histochem. J.* **3** 233–240.

Pearce, R. B. (1986) Chlorantine Fast Green BLL as a stain for callose in oak phloem. *Stain Technol.* **61** 47–50.

Pearse, A. G. E. (1960) *Histochemistry: theoretical and applied*, 1st edn. Boston, Little, Brown.

Pearse, A. G. E. (1968/1972) *Histochemistry: theoretical and applied*, Vols. 1/2, 3rd edn. Edinburgh, Churchill-Livingstone.

Perls, M. (1867) Nachweis von Eisenoxyd in geweissen Pigmentation. *Virchows Arch. pathol. anat. physiol. klin. Med.* **39** 42–48.

Peters, A. (1955) Experiments on the mechanism of silver staining. Part I. Impregnation. Part II. Development. *Quart. J. micr. Sci.* **96** 84–102 and 103–115.

Polak, S. M. & Noorden, S. (1983) *Immunocytochemistry. Practical applications in pathology and biology*. Bristol, John Wright.

Proctor, G. B., Bayliss High, O., Horobin, R. W. & Self, T. (1984) Staining hydrophobic substrates, a structure–staining analysis. *Proc. roy. micro. Soc.* **19** 34–35.

Puchtler, H., Meloan, S. N. & Terry, M. S. (1969) On the history and mechanism of Alizarin and Alizarin Red S stains for calcium. *J. Histochem. Cytochem.* **17** 110–124.

Sandritter, W., Kiefer, G. & Rick, W. (1966) Gallocyanin–chrome alum. In *Introduction to quantitative cytochemistry* (ed. Wied, G. L.). New York, Academic Press.

Scott, J. E. (1967) On the mechanism of the methyl green–pyronin stain for nucleic acids. *Histochemie* **9** 30–47.

Scott, J. E. (1972) Histochemistry of Alcian Blue. III. The molecular biological basis of staining by Alcian Blue 8GX and analogous phthalocyanins. *Histochemie* **32** 191–212.

Scott, J. E. (1973) Affinity, competition and specific interactions in the biochemistry and histochemistry of polyelectrolytes. *Biochem. Soc. Trans.* **1** 787–806.

Scott, J. E. (1980) The molecular biology of histochemical staining by cationic phthalocyanin dyes. The design of replacements for Alcian Blue. *J. Microsc.* **119** 373–381.

Scott, J. E. & Harbinson, J. (1968) Periodic oxidation of acid polysaccharides. Inhibition by the electrostatic field of the substrate. *Histochemie* **14** 215–220.

Shoobridge, M. (1983) A new principle in polychrome staining. *Stain Technol.* **58** 245–257.

Sippel, T. O. (1978) The histochemistry of thiols and disulphides. III. Staining patterns in rat tissues. *Histochem. J.* **10** 597–609.

Spiers, R. L. (1970) The reaction between some dyes and synthetic hydroxyapatite. 1. The uptake of dyes in relation to their structure. *Histochem. J.* **2** 45–66.

Stoward, P. J. & Polak, J. M. (1981) *Histochemistry: the widening horizons of its application in the biomedical sciences*. Chichester, John Wiley.

Sumner, A. T. (1977) Suppression of Quinacrine banding of human chromosomes by mounting in organic media. *Chromosoma* **64** 337–342.

Sumner, A. T., Carothers, A. D. & Rutovitz, D. (1981) The distribution of quinacrine on chromosomes as determined by X-ray microanalysis. *Chromosoma* **82** 717–734.

Tas, J., Mendelson, D. & Noorden, C. J. F. (1983) Cuprolinic Blue: a specific dye for single-stranded RNA in the presence of magnesium chloride. I. Fundamental aspects. *Histochemical J.* **15** 801–813.

Thompson, S. W. (1966) *Selected histochemical and histopathological methods*. Springfield, Thomas.

Thompson, S. W. & Luna, L. (1978) *An atlas of artifacts: encountered in the preparation of microscopic tissue sections*. Springfield, Thomas.

Uchida, I. A. & Lin, C. C. (1974) Quinacrine fluorescent patterns. In *Human chromosome methodology* (ed. Yuni, J. J.). New York, Academic Press, Chapter 3.

Underwood, J. C. E. (1987) *Introduction to biopsy interpretation and surgical pathology*. Berlin, Springer.

Vacca, L. L. (1984) *Laboratory manual of histochemistry*. New York, Raven Press.

Velikan, C. & Velikan D. (1972) Silver impregnation. In *Techniques of biochemical and biophysical morphology* (eds Glick, D., Rosenblaum, R. M.). New York, John Wiley, Vol. 1, pp. 143–190.

Whistler, R. L. & Daniel, J. R. (1979) Starch. In *Kirk–Othmer encyclopedia of chemical technology*. 3rd edn. New York, John Wiley, Vol. 12, pp. 492–507.

Wittekind, D. H. & Gehring, T. (1985) On the nature of the Romanowsky–Giemsa effect. I. Model experiments on the specificity of Azur B–Eosin Y stain as compared with other thiazine dye–Eosin Y combinations. *Histochem. J.* **17** 263–289.

Index

QH
613
H67
1988

SUNY FREDONIA
3 0243 00270641 4